D1719299

Manufacturing of Concrete Products and Precast Elements

VLB record

Helmut Kuch / Jörg-Henry Schwabe / Ulrich Palzer
Manufacturing of Concrete Products and Precast Elements
Processes and Equipment
Verlag Bau+Technik GmbH, 2010

ISBN 978-3-7640-0519-1

Produced by: Verlag Bau+Technik GmbH,
P.O. Box 12 01 10, 40601 Düsseldorf
www.verlagbt.de

Originally published in German in 2009 as:
Herstellung von Betonwaren und Betonfertigteilen
Translated into English by Steffen Walter and Gillian Scheibelein

Printed by: B.O.S.S Druck und Medien GmbH, 47561 Goch

Manufacturing of Concrete Products and Precast Elements

Processes and Equipment

Dozent Dr.-Ing. habil. Helmut Kuch
Prof. Dr.-Ing. Jörg-Henry Schwabe
Dr.-Ing. Ulrich Palzer

Presented by:

Table of Contents

Foreword

Concrete is one of the most important building materials of our times. Concrete products and precast elements that are prefabricated on an industrial scale fully utilise the performance potential of concrete whilst offering major benefits with regard to the construction process. The flexible use of prefabricated concrete products results in a continuously increasing diversity with respect to

- fresh concrete mix designs and properties,
- external geometry and design,
- surface finishes in terms of colour and design and
- characteristics of the finished product (quality).

These factors impose corresponding requirements on both the manufacturers of the associated production equipment and its operators, i.e. precast plants.

The main objective is to implement a flexible production system with respect to all four components of the production process, i.e.

- material-related aspects,
- technological processes,
- technical equipment and
- characteristics of the finished product (quality).

These components need to be carefully considered and evaluated to ensure that the concrete products and precast elements are manufactured to the required quality standards.

The relevant literature does not include any comprehensive discussions of these relationships to date.

This book is based not only on the authors' many years of experience gained in the field of precast technology at the Bauhaus University of Weimar and at the Institut für Fertigteiltechnik und Fertigbau Weimar e. V. (Weimar Institute for Precast Technology and Construction), but also on their close ties to the industry.

The authors' aim was to select state-of-the-art testing and calculation methods from neighbouring disciplines and apply them to precast technology. This includes, for instance, modelling and simulation of the workability behaviour of mixes, application of the latest advancements in machine dynamics to the design and engineering of production equipment, and the use of state-of-the-art measuring and automation technology for quality control purposes.
In the English translation, the system of mathematical symbols and designations used in the German version was intentionally retained. The same applies to the metric units of measurement used for physical parameters.

We thank all those who contributed to the publication of this book, in particular Prof. Dr.-Ing. habil. Dieter Kaysser, Dr.-Ing. Steffen Mothes and Dipl.-Phys. Günter Becker for their active involvement.

We are also grateful to numerous companies for providing photographs.

The authors particularly appreciate the assistance of the following industry partners in supplying useful information and images during the preparation of this book:

Avermann, Osnabrück; BETA Maschinenbau, Heringen; BHS Sonthofen; Dreßler Bau, Stockstadt; EBAWE, Eilenburg; Eirich, Hardheim; Elematic, Nidda; Fritz Hermann, Kleinhelmsdorf; Hess, Burbach-Wahlbach; HOWAL, Ettlingen; Knauer Engineering, Geretsried; KOBRA, Lengenfeld; Hawkeye Pedershaab, Bronderslev, Denmark; Liebherr-Mischtechnik, Bad Schussenried; NUSPL BETONWERKSEINRICHTUNGEN, Karlsruhe-Neureut; PRAEFA, Neubrandenburg; Prinzing, Blaubeuren; Rampf, Allmendingen; REKERS, Spelle; Ruf, Willburgstetten; Schindler, Regensburg; Schlosser-Pfeiffer, Aarbergen; Schlüsselbauer, Gaspoltshofen, Austria; Sommer, Altheim; Technoplan, Seyda; Vollert, Weinsberg; Wacker, Munich; Weckenmann, Dormettingen; Weiler, Bingen; Wiggert, Karlsruhe; ZENITH, Neunkirchen

Our special thanks go to the following individuals who supported us in many ways in designing and publishing this book:

Heike Becker
Dipl.-Ing. Jens Biehl
Dipl.-Ing. Frank Bombien
Dipl.-Ing. Tobias Grütze
Dr.-Ing. Barbara Janorschke
Dipl.-Ing. Jürgen Martin
Kerstin Meyer
Dr.-Ing. Simone Palzer
Dipl.-Ing. Kerstin Schalling
Dipl.-Ing. Christina Volland
Dipl.-Ing. Markus Walter

The authors
Weimar, August 2010

Introduction

Building with state-of-the-art precast reinforced concrete construction evolved into an industrial construction method only over the last 60 years or so. The first attempts to erect buildings using structural elements made of precast reinforced concrete were made at the turn of the 20th century. Examples include the casino in Biarritz (Coignet) in 1891 and prefabricated gatekeepers' houses (Hennebique, Züblin) in 1896. This trend continued across Europe and in the United States during the first half of the last century, and precast technology saw its actual breakthrough after World War II. The huge demand for housing confronted the construction industry with an enormous amount of building work. During this period, the systems developed by the French (e.g. Camus, Estiot) and Scandinavians (e.g. Larsson, Nielsen) provided the key momentum towards large-panel construction. The increasing lack of skilled workers shifted the emphasis to factory production and resulted in the breakthrough of precast products. In addition to systems for industrialised housing construction, the increase in related education and training programmes led to the full emergence of skeleton construction based on structural framework using columns, beams and wide-span floor slabs. For both industrial and sports facilities construction, standardised product ranges were developed that included precast columns, prestressed double-T beams and purlins or shed roofs.

Parallel to these processes, other concrete products were developed for the associated infrastructural facilities above and below ground.

Prefabrication of precast elements and the virtually countless variety of small concrete products require the use of appropriate production equipment. The German building materials machinery sector made a particularly significant contribution to respond to this need, which is why German equipment manufacturers are global market leaders today. A major factor that had to be taken into account were ongoing developments in the materials field, which have a significant impact on precast technology.

About 25 years ago, concrete was still a conventional ternary mixture comprising cement, water and aggregate. In addition to these three main constituents, it now contains additives (e.g. workability agents or retarders) and additives (e.g. coal fly ash). This trend enabled significant widening of the performance range of concrete. Modern product ranges include high-strength, fibre-reinforced and self-compacting grades.

Further material developments in the precast sector include optimisation of lightweight concrete by adding suitable lightweight aggregates (e.g. expanded clay, shale or glass, pumice, lava, lightweight sand, perlite) or using artificially introduced pores or foams. New areas of application are opening up for high-performance concretes containing fine-grained aggregates and textile reinforcement in combination with new design and placement principles. Chemical additives play a crucial role in making the material more sustainable, enabling more slender elements, and utilising concrete in a specific and economical manner.

The current state of the art also includes reinforcing fibres that are added to enhance the viscosity, strength and crack resistance of concrete, which would otherwise remain brittle. The use of textile mesh reinforcement or various fibres (carbon, glass, basalt, polymers) is fostering the development of new concrete grades with a better performance in terms of their impermeability, structural design and strength, as well as their material and surface qualities.

Strengthening concrete with fibreglass-reinforced plastics has opened up new markets on account of their new material quality parameters (e.g. corrosion resistance, electrical insulation, non-magnetic properties and resistance to chemical attack).

New developments in the concrete and precast industry are driven not only by the rising costs of energy and raw materials, but also by increasingly stringent product quality standards with respect to thermal insulation, durability and resistance of the products to environmental effects and other characteristics that depend on their intended use.

The design options for concrete products will extend their range of application. Such options include various concrete surface finishes that are achieved by washing, fine washing, acid washing, blasting, flame cleaning, grinding and polishing, by applying stonemasonry techniques, by creating coloured surfaces as a result of adding various cements, mineral aggregates and pigments, by painting or by photo-engraving.

This diverse range of design options for the concrete products requires suitable manufacturing processes and equipment.

These aspects are the focus of this book. It has been written for everyone involved in the production of prefabricated concrete products, including:

- manufacturers of production equipment,
- users and operators of such equipment, i.e. concrete and precast plants,
- students enrolled in related degree courses and advanced training,
- researchers and developers of processes and equipment in the field of precast technology.

The current situation is characterised on the one hand by increasingly diverse concrete products, and on the other, by the great degree of variety and numerous control options offered by commercially available equipment.

The aim is to develop a flexible manufacturing process for prefabricated concrete products that conform to a high quality standard. This necessitates clarification of the complex relationships between the various components of the concrete production process, namely:

- material-related aspects,
- technological processes,

- technical equipment and
- characteristics of the finished product (quality).

In many cases, however, these factors are still being dealt with on an empirical basis.

Mastery of these complex processes requires that all parties involved must cooperate as closely as possible. This applies, in particular, to the manufacturers and operators of the production equipment. To achieve this goal, they should have a sound knowledge of the basic underlying principles and interactions.

From their many years of experience gained in close collaboration with industrial partners, the authors concluded that this was exactly where a real gap existed in the literature on precast technology, which is why they decided to write this book.

Chapter 1 outlines the basic principles required to understand the interactions referred to above.

The process for manufacturing concrete products is first described on the basis of the process elements, process layout and process flow. The processing behaviour of concrete is described with particular attention paid to moulding and compaction of the concrete mix. The associated processing parameters are defined.

This chapter also describes the raw materials used to produce the concrete mix whilst also looking at the concrete mix design in greater detail. The evolution from a ternary mixture to the current quinary system is also discussed. The empirical solutions commonly applied in the past will be increasingly replaced by process optimisation and simulation exercises that take account of the properties of the concrete mix, fresh and hardened concrete as well as their testing.

The fundamentals of the products are outlined starting with a clear definition of the concrete products and product groups whose manufacture is described in subsequent chapters. This is followed by a discussion of the requirements for the product properties and a description of the associated testing methods.

In the chapter describing the basic aspects of the equipment, reference is first made to the various types of vibration equipment, which is crucial for the manufacture of concrete products.

The current situation with regard to modelling and simulation of the workability behaviour of mixes is then described. This option to evaluate processing work steps in conjunction with laboratory-, pilot- and industrial-scale testing is becoming increasingly popular. The development of the associated hardware and software will strengthen this trend. The application of these principles is demonstrated in Chapter 2: the processes and equipment required to produce the concrete mix are described for all prefabricated concrete products.

The same applies to the dynamic modelling and simulation of production equipment. Modelling of equipment using

- multi-body systems and
- the Finite Element Method (FEM)

can be used to investigate motion processes as well as stresses generated by dynamic loading. The application of these simulation methods is then described along with the individual equipment components.

The processes and equipment to manufacture precast concrete products are then discussed for the individual product groups:

- small concrete products,
- concrete pipes and manholes,
- precast elements.

The characteristics of the final product are of crucial importance, which is why in-process quality control is becoming increasingly popular. Implementation of a quality control system requires state-of-the-art measuring and automation technology, which is also discussed in this book.
Also addressed are issues associated with appropriate measures for reducing noise and vibration during the manufacture of precast products.

1 Basic Principles

1.1 Process Fundamentals

1.1.1 Production Process

The production process to manufacture concrete products can also be considered a system, just like any other process. The schematic representation shown in Fig. 1.1 indicates the system boundaries of the production process.

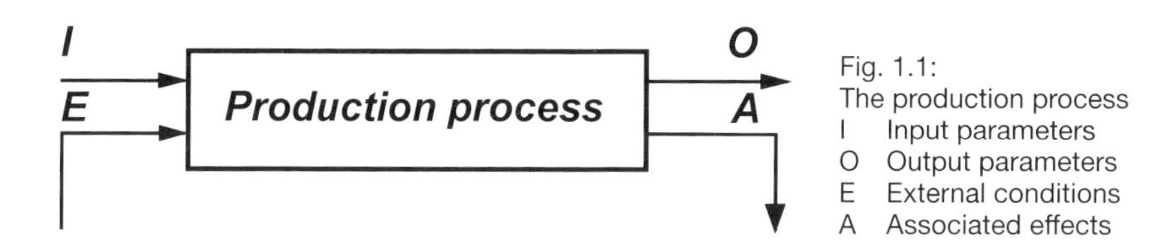

Fig. 1.1:
The production process
I Input parameters
O Output parameters
E External conditions
A Associated effects

As is the case with any system, the basic characteristics of this production process are its function and structure. The function of the production process is the conversion of certain input parameters (e.g. material, energy or information) into the associated output parameters (e.g. semi-finished and finished products). The structure of the production process serves to fulfil the function and includes a set of elements that are interlinked by particular relationships.

The production process is subject to certain conditions that must be considered during the planning, preparation and execution stages. These are:

- on the input side, conditions that restrict the degree to which the function can be fulfilled; these include environmental factors, available equipment and conditions of supply.
- on the output side, conditions associated with the fulfilment of the function; these include emissions and by-products generated by the process.

1.1.2 Components Determining the Structure of the Production Process

The process to manufacture concrete products is a complex, dynamic system made up of technical and organisational elements.

Process elements are basic processes or workflows that can no longer be sub-divided from a macro-technological perspective. These process elements are linked by temporal, spatial and quantitative relationships that are determined by the process function.

These relationships govern the process layout and flow with respect to both space and time.

Therefore, the following parameters need to be determined to describe the production process fully:

- process elements
- process layout
- process flow

1.1.2.1 Process elements

Like any other process, the basic operation, as a process element, has both a function and a structure.The function of the basic operation is a fundamental change in the state of the target object towards the final product and aims to achieve a certain intermediate state.

All existing objects can be assigned to one of the following main categories:

material - energy - information

They are modified by basic operations of the various types of change, all of which can be assigned to the following categories:

production - transport - storage

Depending on the relevant type of change, the basic operations are elements that determine the production process and can most generally be described, from a functional point of view, as:

a) production elements
b) transport elements
c) storage elements

With regard to the overall production process, the characteristics of its elements also form the basis for its constituents:

1 Manufacturing technology and organisation
 1.1 Production technology and organisation
 1.2 Transport technology and organisation
 1.3 Storage technology and organisation.

2 Manufacturing-related technology and organisation
 2.1 Supply and disposal technology and organisation
 2.2 Maintenance technology and organisation
 2.3 Safety and security technology and organisation
 2.4 Control technology and organisation.

The structure of the basic operations forms part of the technological microstructure (Fig. 1.2).

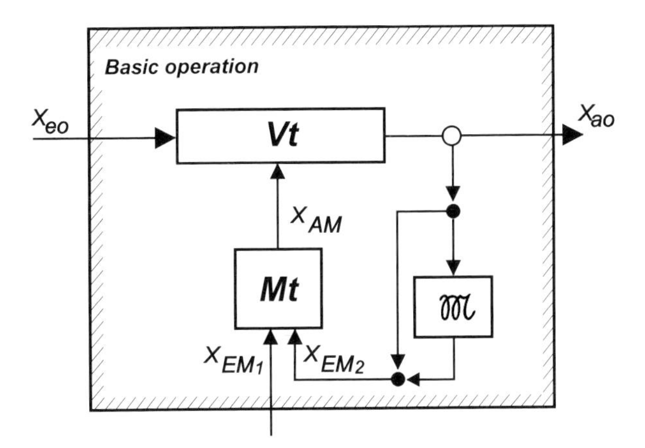

Fig. 1.2:
Structure of the basic operation

Structural elements thus include:

- the object of change (X_e, X_a),
- the technological method (Vt),
- technical means (Mt),
- human workforce ($\mathfrak{M}$).

In the basic operation, a human being uses a technical means to affect the object of change directly or indirectly, thus modifying it with a certain aim or purpose. The technological method governs the basic way in which this proceeds. Technological methods thus represent the approach usually applied in practice to implement scientific effects and to thus modify the object in accordance with the intended purpose. The technological method is not an object itself, it is inherent to the technical means that fulfils its function within the technological process.

The technical means represents a technical object that can be considered a system, i.e. a technical entity (technical equipment). The function of the technical means is to implement one or several technological methods within the technological process.

In accordance with this function, it is useful to classify these means analogously to the functional relationships between the basic operations (i.e. according to the type of change):

- production means
- transport means
- storage means

Furthermore, the technical means may also be categorised according to the object of change:

- material-related means
- energy-related means
- information-related means

The set of material-related means comprises all technical means that serve to change the state of materials in the most general sense. These include all pieces of equipment (such as machines, apparatus, devices and systems) that are used to manufacture products from the materials.

Energy-related means comprises all technical means that convert or transform energy, such as drive motors, steam generators, transformers or energy distribution systems.

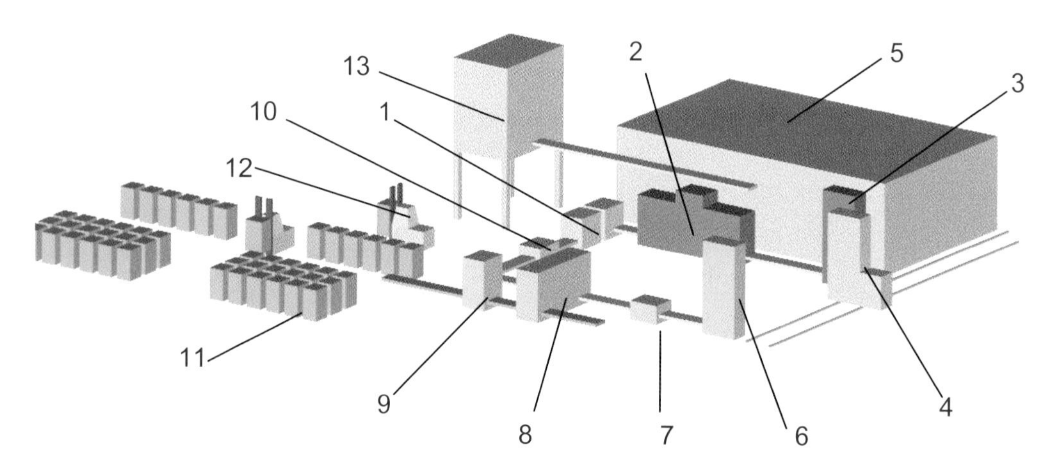

Fig. 1.3: Technological line (plant) to manufacture concrete products

1 Pallet buffer
2 Block machine
3 Elevator
4 Transfer car with top finger car
5 High-bay rack/curing chamber
6 Lowerator
7 Quality control
8 Re-arranging and stacking unit
9 Strapping system
10 Cleaning, turning and stacking of pallets
11 Storage of products ready for dispatch
12 Transport equipment
13 Mix processing and feed

Information-related means comprise all technical means that serve to process information. These include, for instance, IT systems, signalling installations, measuring equipment as well as weighing and batching units.

The coupled set of technical means used in the production process represents the production line, which is an overall entity (technological line) and is also a prerequisite to carry out the production process (Fig. 1.3). The technical means are thus at the very heart of the various processes.

1.1.2.2 Relationships between process elements

Within the technological process, certain relationships exist between the process elements that are determined in space and time. As a result, the set of relationships between the process elements represents the spatial and temporal organisation of the technological process, i.e. the process layout and flow. Both sides of the structure are governed by the following underlying conditions that must be met by an appropriately designed structure:

1 Fulfilment of the function
  1.1 Ensuring both quality and quantity of products
  1.2 Implementing the required functional sequence

2 Process efficiency
  2.1 High reliability
  2.2 Lowest possible outlay for process installation and implementation

3 People-driven nature of the process
  3.1 Best possible working conditions
  3.2 Lowest possible emissions.

Process layout and flow comprise the set of arrangements and couplings between process elements.

These arrangements determine the position of process elements in space and time.

The spatial arrangement is thus defined by the allocation of the process elements to the required functional sequence and the associated flow of materials, as well as by the options that exist with respect to the set-up and positioning of the technical means. The arrangement within the production space depends on the functional and geometrical/structural characteristics of the technical means, on the requirements for their assembly, operation and maintenance, and on the characteristics of the production space.

The temporal arrangement of the process elements is determined by the required functional sequence and by factors associated with the output parameters and work scheduling.

Couplings are the links that permit transfer of the object of change (material, energy, information) between process elements.

The overall set of couplings comprises:

- spatial-geometric couplings
- temporal couplings and
- quantitative couplings.

Certain compatibility conditions must be met in order to fulfil the coupling function. To achieve compatibility, the output variables of the preceding operation must correspond to the input variables of the subsequent operation with respect to space, time and quantity. If this condition is not met, the operations cannot be coupled. In such a case, either a modification of the elements to be coupled or the integration of additional elements is required.

In this model, a spatial coupling refers to a spatial-geometrical relationship between process elements. This requires geometrical compatibility at the spatial points where objects of change are transferred. For this purpose, the three-dimensional coordinates of the boundaries of the process elements (the technical means) are aligned with each other in such a way that the objects of change can be transferred. A spatial coupling must fulfil the following conditions:

- transfer of the object of change must be ensured
- mobile technical means must have enough space to manoeuvre
- sufficient space must be provided for assembly, repair and maintenance.

This leads to specific coupling distances (Fig. 1.4).

Fig. 1.4: Spatial coupling of process elements: concrete mix spreader above the pallet

Fig. 1.5: Serial process: pallet circulation

A temporal coupling refers to the alignment of process times of the various process elements. Two process categories can be distinguished with respect to their temporal characteristics:

- serial processes
- parallel processes

Serial processes require that a process element must have been completed before the following element can commence. In this case, the time gap amounts to $t_{1,2} \geq 0$ (Fig. 1.5).

Parallel processes require that all parallel processes involved must have been completed at the lateral nodes so that they can be merged into a common process. In this model, the co-determinative processes must be adjusted to the determinative process (Fig. 1.6).

The following factors are relevant to quantitative couplings:

- The majority of process elements that are coupled to create a production process provide varying capacities, which results in different mass flows.
- With respect to their capacities, technical means of a single type are mainly composed of discrete increments.
- The required mass flows can be achieved either by a large-capacity process element or by several process elements whereby each of these elements provides a lower capacity.
- Process elements that include various types of flow (i.e. continuous vs. discontinuous) may have to be coupled within a single production process.

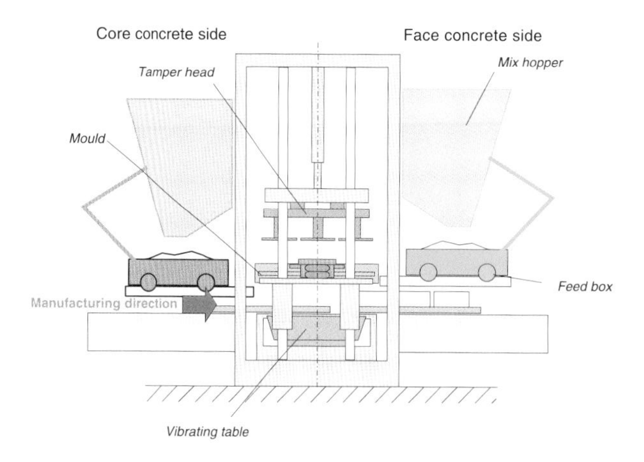

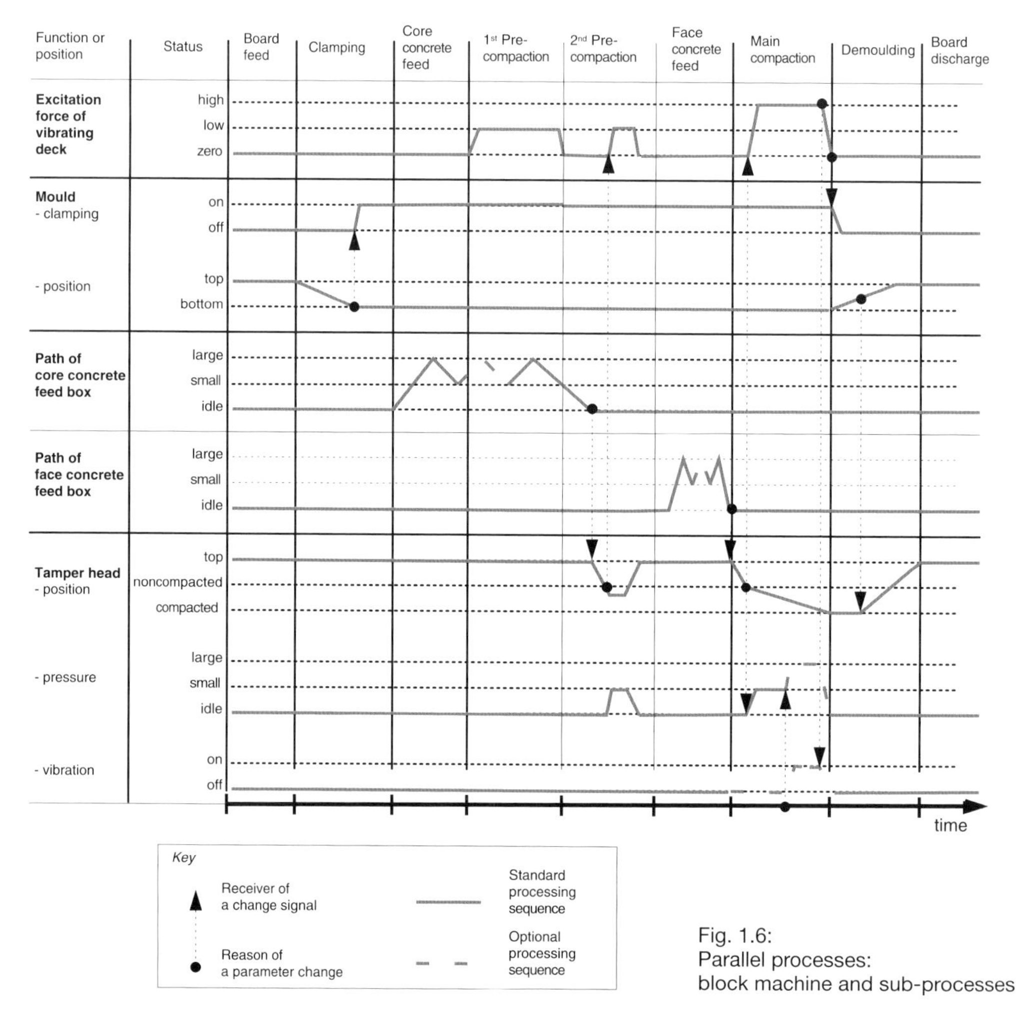

Fig. 1.6:
Parallel processes:
block machine and sub-processes

The following compatibility condition applies to the quantitative coupling of two consecutive process elements:

$$\dot{M}_1 = \dot{M}_2 \tag{1.1}$$

$\dot{M}$ mass flow

$$\dot{M} = \frac{dM}{dt} \tag{1.2}$$

Technical or organisational adjustments need to be made if the mass flow lines diverge ($\dot{M}_1 > \dot{M}_2$) or converge. ($\dot{M}_1 < \dot{M}_2$)

Options for technical adjustments are:

- modification of factors that determine the capacity of the process elements to be coupled by changing the material quantity or the production or conveying speed,
- integration of additional intermediate or parallel elements. Process elements to be integrated as intermediate elements mainly include storage elements that are introduced for compensation purposes and which put a certain number of objects of change on hold for a defined period (Fig. 1.7).

A parallel arrangement is required if there are process elements with varying flow increments. In this case, a single, larger-flow element is coupled to several elements with smaller flows in such a way that an alignment is achieved.

Fig. 1.7:
Insertion of storage elements: storage system for baseboards

1.1.2.3 Process layout and flow

The process layout determines the spatial structure.

The spatial structure refers to the three-dimensional arrangement and coupling of the process elements. It represents the spatial organisation of the technological process and thus of the production line as the entity that comprises all technical means [1.1]. Its configuration can be varied according to the following types of spatial organisation:

- basic types of arrangement
- types of motion

a) Basic types of arrangement
Basic types of arrangement are distinguished according to the process- or product-driven nature of the spatial arrangement.

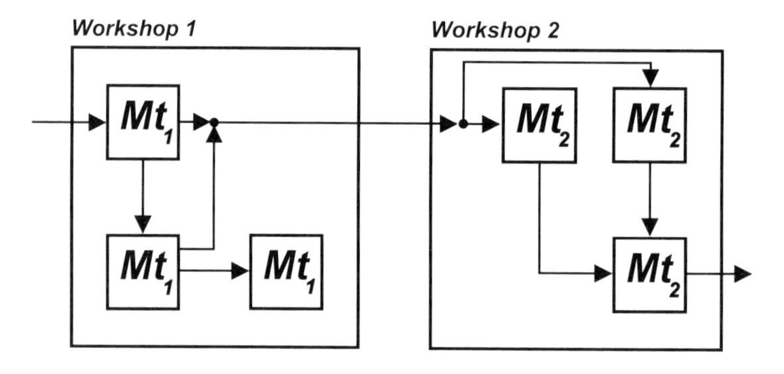

Fig. 1.8:
Process-driven arrangement

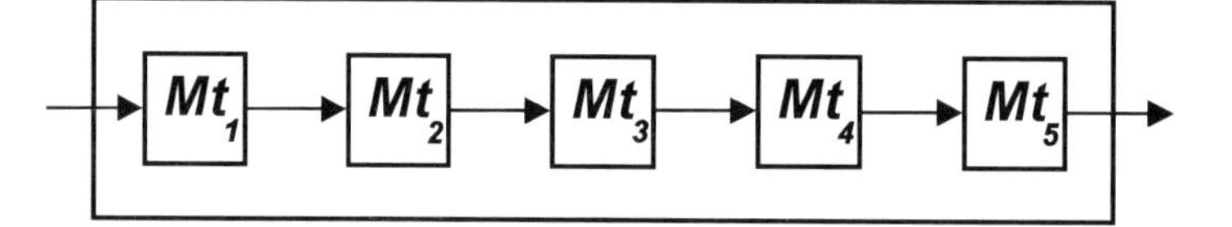

Fig. 1.9:
Product-driven arrangement

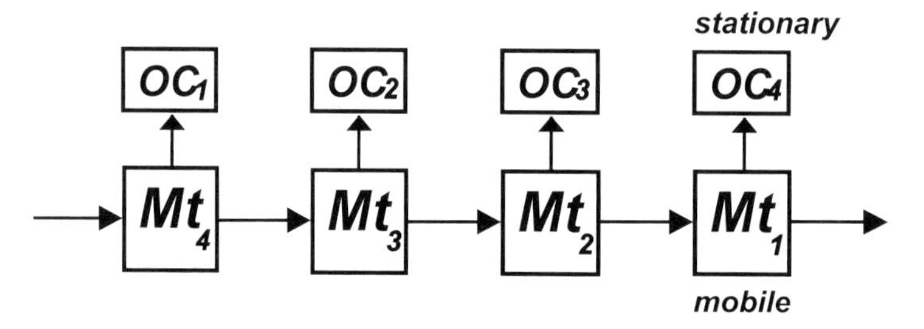

Fig. 1.10:
Stationary production

Process-driven arrangement *(process principle):*
Technical means that implement identical processes are grouped together in a spatial arrangement and treat various types of objects of change (Fig. 1.8).

Product-driven arrangement *(product principle):*
Technical means that implement different processes are grouped together in a spatial arrangement according to the work sequence required for a certain type of object of change (Fig. 1.9).

b) Types of motion
Types of motion can be distinguished according to the state of motion between objects of change and technical means:

Stationary production
The objects of change (OC) remain at the same manufacturing station during the determinative basic operations. The technical means (Mt) are mobile. They are moved towards the object of change, where they act on it, and are then moved to the next manufacturing station (Fig. 1.10).

The principle of stationary production is used by a number of different systems, of which the following are of particular relevance to the production of wall and structural framework elements:

- single-mould systems
- battery mould systems
- continuous moulding systems
- extrusion systems
- prestressing line systems

Fig. 1.11:
Battery mould

Fig. 1.11 shows a battery mould.

Sequential production

The objects of change (OC) move from one manufacturing station to the next. The technical means Mt are stationary (Fig. 1.12).

One or more work steps are carried out at each of the stations (manufacturing units), which is why these work steps run parallel to each other [1.2]. Fig. 1.13 shows a typical example of the carousel manufacturing principle: a pallet circulation system. Block machines used to manufacture concrete products are another example of this manufacturing principle.

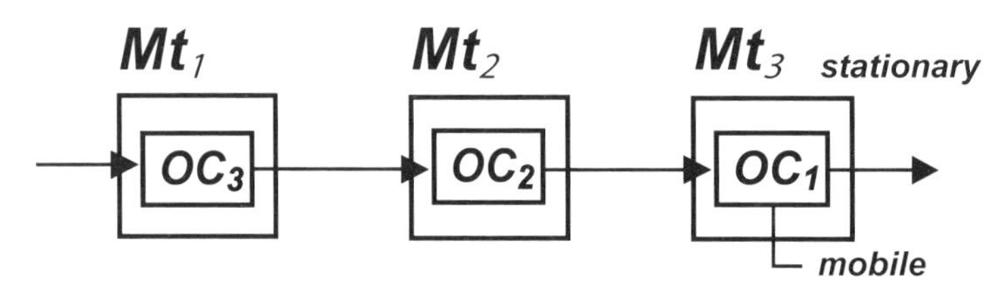

Fig. 1.12:
Sequential production

Fig. 1.13:
Pallet circulation system

1.1.3 Processes for the Industrial Manufacturing of Concrete Products

Concrete products include durable goods made of concrete, reinforced concrete and prestressed concrete [1.3].

Finished concrete is made in the following stages:
concrete mix → fresh concrete → hardened concrete.

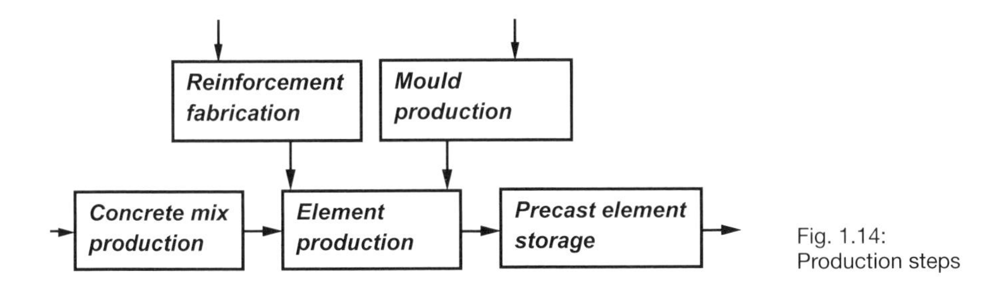

Fig. 1.14:
Production steps

In accordance with these stages, concrete products are manufactured in the following sequence (Fig. 1.14):

- production of the concrete mix
- fabrication of the reinforcement
- production of moulds and formwork
- production of concrete elements
- finishing and completion (partly integrated in element production)
- storage of precast elements and products

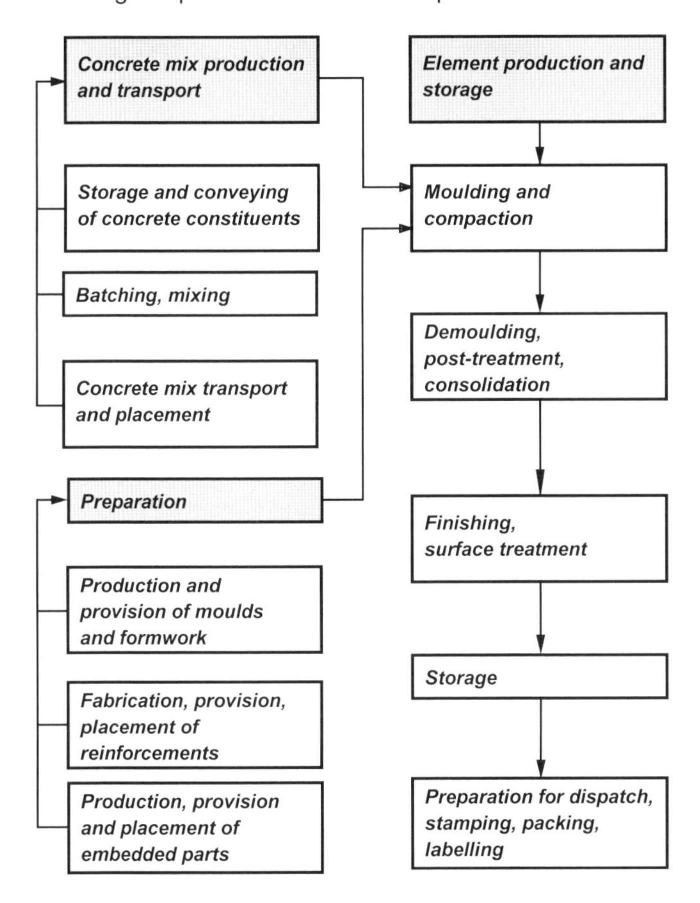

Fig. 1.15:
Sub-processes in the manufacture of concrete products

In this workflow, the production of concrete elements is the main process to shape and manufacture the concrete products. Fig. 1.15 shows the work steps that are required for this purpose.

The steps of mix production as well as fabrication of reinforcements, moulds and formwork may be allocated to one or several element production processes. They may also be located outside the boundaries of the factory; however, this would increase outlay for organisation and transportation.

1.1.4 Processing Behaviour of Concrete

1.1.4.1 Classification of processing behaviour

The manufacture of concrete products requires a number of changes in the state of the material to achieve a defined manufactured state at each of these stages. During these changes in the state or condition, which are brought about by the intentional action of the technical means, the respective object (i.e. concrete constituents or concrete at each of its stages) exhibits a certain behaviour. In other words, this constitutes the reaction of the material to the action of the technical means. The processing behaviour is thus process-driven. In accordance with the main classes defined for the types of change, main processing behaviour classes can also be established (Table 1.1).

Table 1.1: Processing behaviour classes

Behaviour during processing	Mixing Placement Compaction Consolidation Post-treatment Finishing
Behaviour during transport	Belt conveying Pipe conveying Bucket conveying Feed and discharge
Behaviour during storage	Bulk properties Container pressure Stacking behaviour

1.1.4.2 Compaction behaviour of the concrete mix

Just like finished concrete, the initial concrete mix is a very versatile material. With respect to its mechanical properties, it takes an intermediate status between a bulk material and a suspension. These mechanical characteristics undergo substantial changes during the compaction process, which thus alters the compaction behaviour.

Compaction is closely related to the moulding behaviour of the concrete mix to produce the concrete product. Moulding and compaction serve to transform the concrete mix into a quasi-solid geometric body of fresh concrete [1.4]. This process creates an artificial stone that has a low initial strength, the so-called green strength.

The aim of the moulding process is to produce an accurately shaped concrete product. The concrete mix is poured into the mould so that it completely fills all the corners and edges. The placement behaviour of the concrete is crucial to achieve this goal and depends on the flow properties of the concrete mix.

For most types of concrete mixes used to manufacture concrete products, natural compaction effects are also utilised to support the placement process. Highly flowable mixes, such as self-compacting concretes (SCCs), show a very good pouring behaviour because any remaining pores are removed by the gravity effect and the motion of the mix during the placement process. As these concretes are already self-compacted, additional compaction is neither necessary nor possible.

Compaction serves to largely eliminate the external porosity of the concrete mix. The reduction in the void volume should lead to higher densities and thus improve the strength and dimensional stability.

Fresh concrete may thus be considered dense if it is largely free of pores.

Concrete can be considered strong if an almost homogeneous body held together by adhesive and cohesive forces was created due to the high packing density of the concrete constituents.

Concrete can be considered dimensionally stable if no significant dimensional changes occur under ambient conditions in both the loaded and the unloaded states.

a) Moulding and compaction methods
Concrete can be compacted by a number of different methods (Fig. 1.16).

Despite numerous attempts to find alternative methods, vibration – alone or in combination with other processes – continues to be the most popular method for moulding and compacting concrete mixes in order to manufacture both concrete products and precast elements [1.5].

Moulding and compaction aims to:

- match the process and equipment parameters to the respective concrete mix and to implement these parameters in the compaction equipment,
- uniformly transfer the required compaction energy into the concrete mix from all points or areas of introduction,
- ensure, by the selection of the appropriate vibration parameters, that the compaction energy in the concrete mix is transferred in such a way that the concrete or precast product has a uniform density throughout.

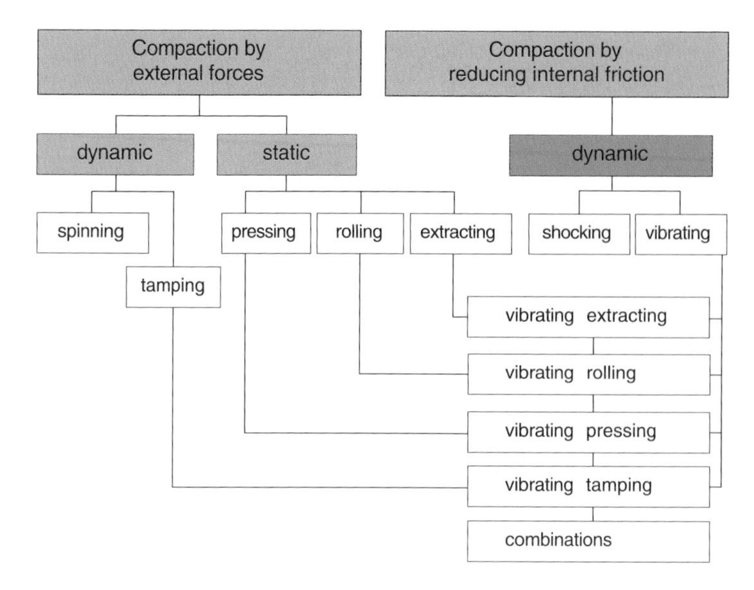

Fig. 1.16:
Compaction methods

The type of action on the concrete mix is a crucial factor that determines the moulding and compaction behaviour. As shown in Fig. 1.17, the various actions can be grouped according to:

- the point of action
- the function of the vibration action
- the intensity of the action
- type, location and number of simultaneous actions
- the phase position of the exciter functions in relation to each other in the case of several simultaneous actions

With respect to the location of the action, and thus its direction, a fundamental distinction can be made between horizontal, vertical and three-dimensional actions.

As regards the function of the vibration action, harmonic and anharmonic modes of excitation can be distinguished. Both directional (counter-acting) and non-directional (circular) exciters can be used to introduce vibration into the concrete. Anharmonic exciter functions can be sub-divided further into periodic and pulsed actions. For instance, a periodic exciter function can be a multi-frequency action that consists of several harmonic components. Pulsed excitation, also known as shock vibration, is generated by shock-like processes. This triggers the inherent oscillation of all system elements capable of vibration, i.e. an entire frequency spectrum.

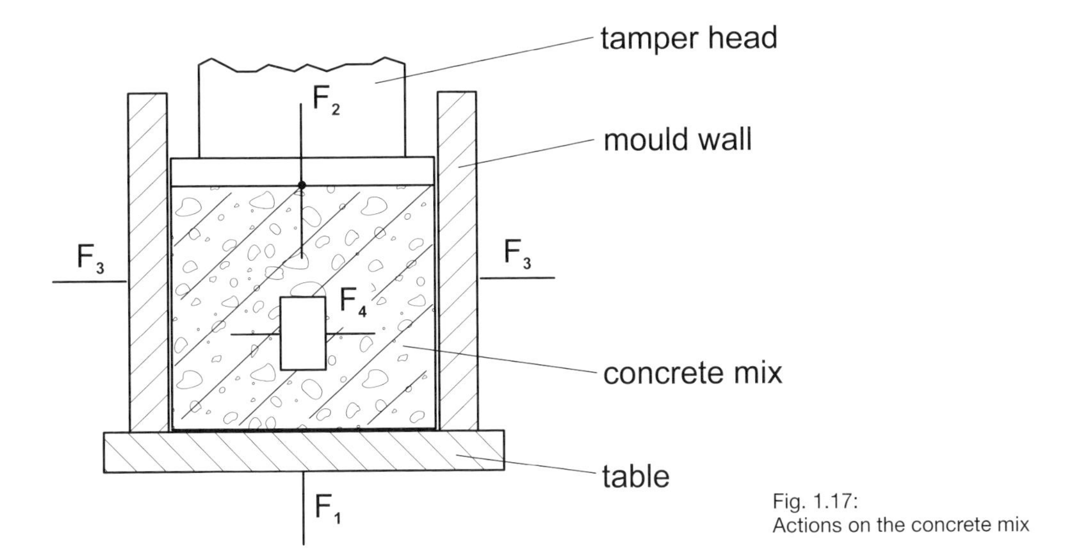

Fig. 1.17:
Actions on the concrete mix

Parameters that characterise the intensity of the action on the concrete mix are discussed in Section 1.1.5.2.

The type of exciter function, the mode of action and number of exciters and, in particular, their phase position in relation to each other have a major influence on the moulding and compaction behaviour of the concrete mix. For example, a phase coincidence of the harmonic vibration components of the vibrating table and tamper head would hardly achieve a good compaction effect.

The crucial factor is the generation of a dynamic pressure gradient between the layers of the mix that enables relative motion of these layers and mutual rotation of the mineral aggregate particles. These requirements must be met by state-of-the-art processes. When producing large-scale precast elements, for example, the low-frequency action on the fresh concrete is complemented by a higher-frequency vertical excitation. In such a set-up, the frequency of the required vibrators is usually controlled via frequency converters.

When producing concrete products from stiff mixes, modern processes often combine vibration with pressing, as in block machines (through the tamper head) or in concrete pipe machines (through a suitable arrangement of packer heads with several level counter-acting rollers).

A special type of action on the concrete mix is created by the use of internal vibrators (Fig. 1.17; excitation force F4), where horizontal, non-directional vibration is introduced when the vibrator comes into direct contact with the mix.

b) The moulding and compaction process

The actual compaction process, from start to finish, can be considered a dynamic process with a gradual transition from one rheological state to the next [1.7]. This concept is illustrated in Fig. 1.18, which is based on investigations carried out by Afanasiev [1.6]. In this model, the entire compaction process is divided into three phases that are described in more detail in [1.7] and [1.8], for example. Each of these three phases represents a compaction stage where, according to [1.6], its rheological state is characterised by the dry friction model, the Bingham model and the Kelvin-Voigt model.

Both duration and delimitation of the individual phases, as well as the associated rheological body models, depend on the type of material mix to be compacted. It can thus be concluded that each concrete mix requires specific process and equipment parameters for the individual phases of its compaction in order to come as close as possible to an ideal compaction state in the shortest possible time [1.8].

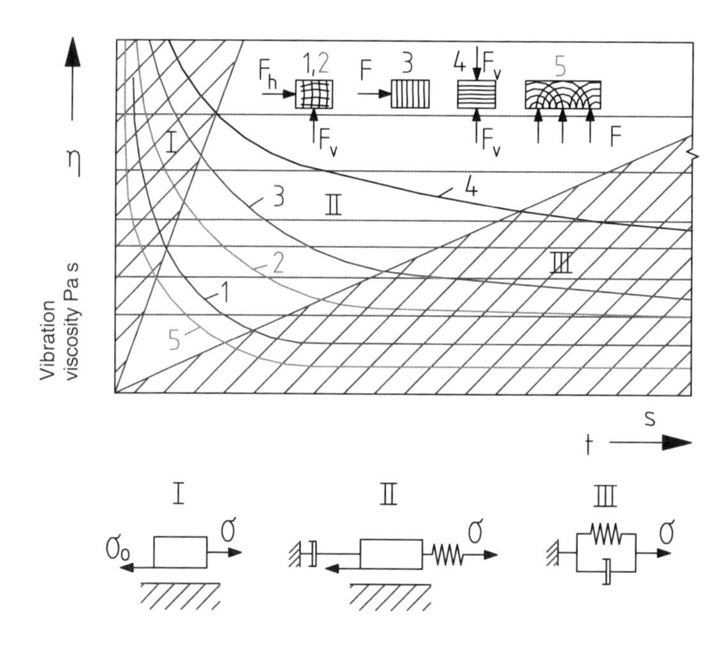

I–III Rheological state of the mixes and their models

F_h horizontal excitation

F_v vertical excitation

Fig. 1.18:
Rheological compactibility curves and corresponding actions on the concrete mix

1.1.4.3 Fundamentals of vibration

a) Kinematics of vibration

The change in a physical parameter over the time period x(t) is termed vibration if the variable x remains finite within the interval considered and if it proceeds from decrease to increase (or vice versa) at least once, i.e. if x(t) does not show monotonic behaviour.

All pieces of equipment and machinery, as well as buildings and their structural components, may be transformed into substitute vibration systems, which means that the actual system (i.e. the framework or structure) is replaced with a vibration model that can be described mathematically for the purpose of describing the vibration behaviour. This model is characterised by the following parameters:

- masses or mass moments of inertia,
- resiliences (springs),
- energy-absorbing elements (dampers, friction pairing).

If an elastically supported mass is deflected from its equilibrium and left undisturbed afterwards, this will result in free oscillation about the (originally stable) equilibrium due to the spring-restoring forces. If damping in the system is neglected, the amplitudes move periodically between two constant extremes. In the simplest case, the amplitudes are pure sine or cosine functions of time (Fig. 1.19). Depending on the specific problem, the variable introduced to characterise the oscillation process represents a path or one of its time derivatives (velocity, acceleration), an angle, a force, a torque, etc. Of the many possible oscillation processes, which will be referred to in more detail subsequently, the harmonic oscillation modes (Fig. 1.19) are most significant because any periodic process can be derived from them by means of a Fourier expansion. Moreover, harmonic functions are applied to many oscillation calculations.

The harmonic oscillation shown in Fig. 1.19 is thus described by a sine function:

$$z = \hat{z} \sin \omega t \tag{1.3}$$

$\hat{z}$ oscillation amplitude
ω angular frequency
t time

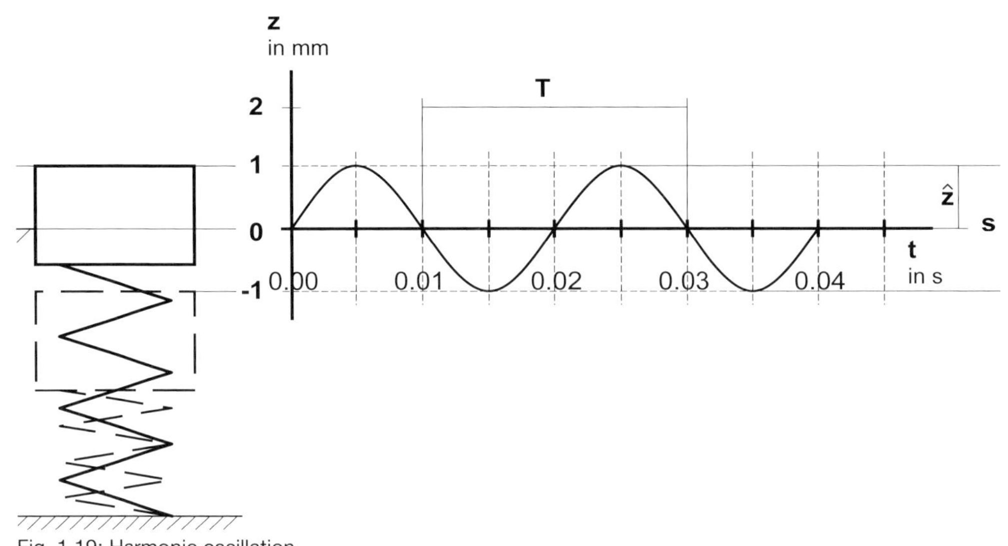

Fig. 1.19: Harmonic oscillation

The time required for a full oscillation is termed a periodic cycle or period of oscillation T:

$$\omega T = 2\pi, \qquad T = 2\pi \frac{1}{\omega} \tag{1.4}$$

Thus the number of oscillations per second, or frequency of oscillation, is calculated as follows:

$$f = \frac{1}{T} = \frac{\omega}{2\pi} \tag{1.5}$$

Both the angular and oscillation frequencies are expressed as s^{-1}. To distinguish between the two, Hz is used as a unit for the frequency of oscillation:

1 Hz = 1 oscillation per second.

A comparison of the angular frequency, or angular velocity, with the speed n expressed in min^{-1} gives:

$$\omega = \frac{2\pi n}{60} \tag{1.6}$$

or

$$f = \frac{n}{60} \tag{1.7}$$

On the basis of Fig. 1.19, the main free oscillation parameters are again summarised in Table 1.2.

Table 1.2: Parameters of a harmonic oscillation

Symbol	Designation	Equation	Unit	Value from Fig. 1.19
T	period of oscillation		s	$T = 0.02$ s
z	vibration displacement		mm	$z(t) = 1\ \text{mm} \cdot \sin(314\ \text{s}^{-1} \cdot t)$
$\hat{z}$	amplitude of vibration displacement		mm	$\hat{z} = 1$ mm
S	range of vibration displacement (peak-peak)	$S = 2\hat{z}$	mm	$S = 2$ mm
f	frequency	$f = \frac{1}{T}$	Hz	$f = 50$ Hz
ω	angular frequency	$\omega = 2\pi f$	s^{-1}	$\omega = 314\ s^{-1}$
$\hat{a}$	acceleration amplitude	$\hat{a} = \omega^2\hat{z}$	ms^{-2}	$\hat{a} = 98.7\ \frac{m}{s^2} \approx 10\ g$

acceleration of gravity $g = 9.81\ \frac{m}{s^2}$

The behaviour over time can be used to make further distinctions beyond purely harmonic oscillation:

- periodic oscillation (Fig. 1.20a) and
- non-periodic oscillation (Fig. 1.20b).

A periodic oscillation exists if the complete cycle is repeated after a certain period (i.e. the period of oscillation T):

$$f(t + T) = f(t) \tag{1.7}$$

In all other cases, a non-periodic oscillation should be assumed. Non-periodic oscillation is often superimposed by shock-like processes and plays a major role in processes and equipment in the precast concrete industry.

For instance, the so-called beat vibration is a common phenomenon seen particularly in vibratory compaction systems (Fig. 1.21). Such a pattern is generated whenever two oscillation cycles with similar frequencies overlap and add up.

b) Forced oscillation

Forced oscillation occurs when a system is constantly excited to trigger vibration, such as the vibrating table of a concrete block machine. Vibration systems are often very

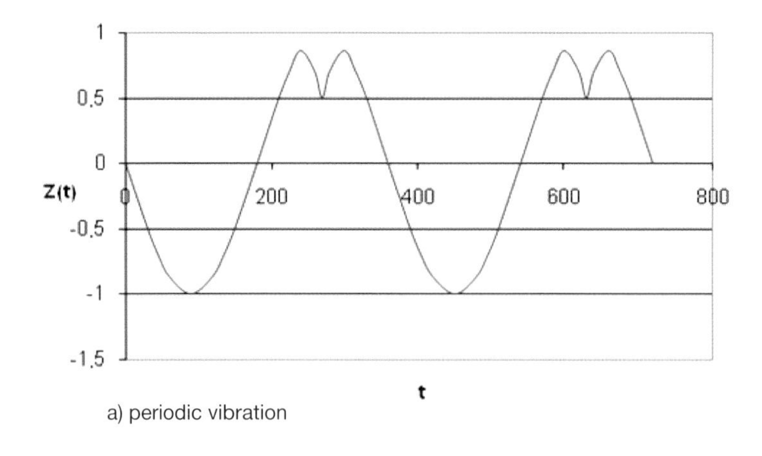

a) periodic vibration

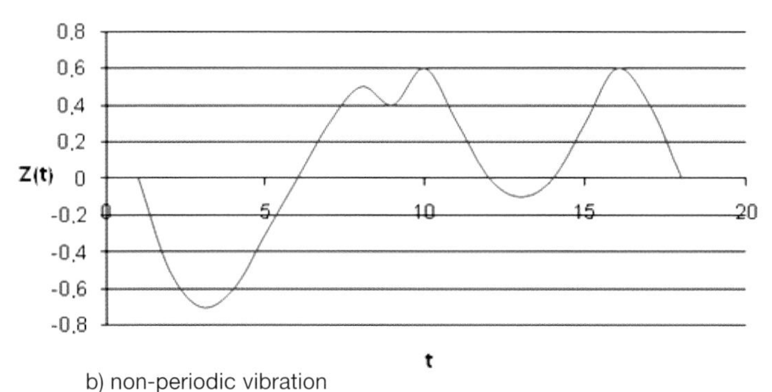

b) non-periodic vibration

Fig. 1.20:
Classification of vibration according to behaviour over time

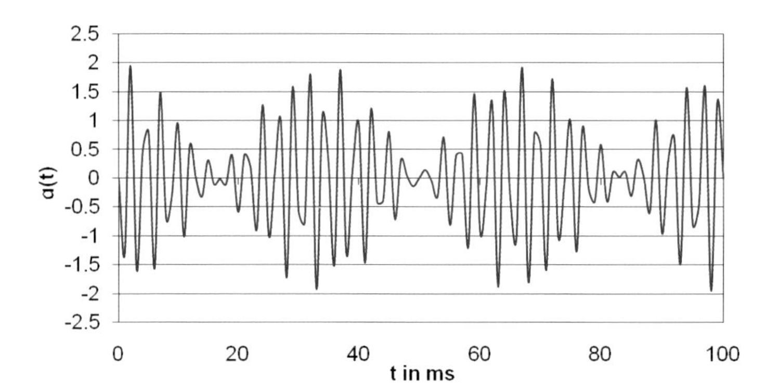

Fig. 1.21:
Beat vibration

complex, which is why a series of assumptions and simplifications needs to be applied in order to be able to carry out the associated calculations. This also means that the theoretical results thus obtained must generally be supported by laboratory, pilot and industrial testing.

It is useful to assume a discrete structure for the initial basic tests: the masses that are assumed to be rigid are connected to each other by zero-mass elastic elements and damping components. The computation model generally includes the following elements:

mass	stores kinetic energy
spring	stores potential energy
damper	dissipates energy
exciter	supplies energy from an external source

Nonetheless, the underlying vibration principles to be used for the design of coupled vibration units can be represented by very simple substitute vibration systems. For this reason, a single-mass system (Fig. 1.22) is assumed in the initial step. The characteristic parameters for this vibration model thus include:

- total mass m
- total spring constant c
- damping coefficient k

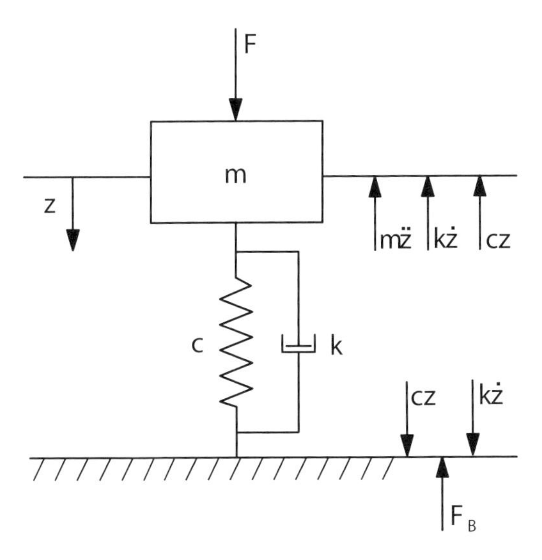

Fig. 1.22:
Substitute vibration system
z coordinate of displacement
m total mass of all vibrating parts
c total spring constant of the elastic support
k damping coefficient
F excitation force

The resulting excitation force is expressed as

$$F = \hat{F} \sin\Omega t. \tag{1.9}$$

If centrifugal excitation force is assumed (Fig. 1.23), it has the following amplitude:

$$\hat{F} = m_u \cdot r_u \cdot \Omega^2 \tag{1.10}$$

where m_u unbalance mass
r_u unbalance radius
Ω angular excitation frequency
t time

d'Alembert's inertial force is determined by Equation (1.11):

$$F_i = m\ddot{z} \tag{1.11}$$

where $\ddot{z} = \frac{d^2z}{dt^2}$ (acceleration)

For the substitute vibration system, this results in the equilibrium of forces shown in Fig. 1.22.

The motion behaviour of the mass can thus be expressed by the following equation of motion:

$$m\ddot{z} + k\dot{z} + cz = F. \tag{1.12}$$

The following steps look exclusively at the stationary (i.e. steady) state. Taking account of Equations (1.9) and (1.10) and introducing the angular eigenfrequency,

$$\omega = \sqrt{\frac{c}{m}} \tag{1.13}$$

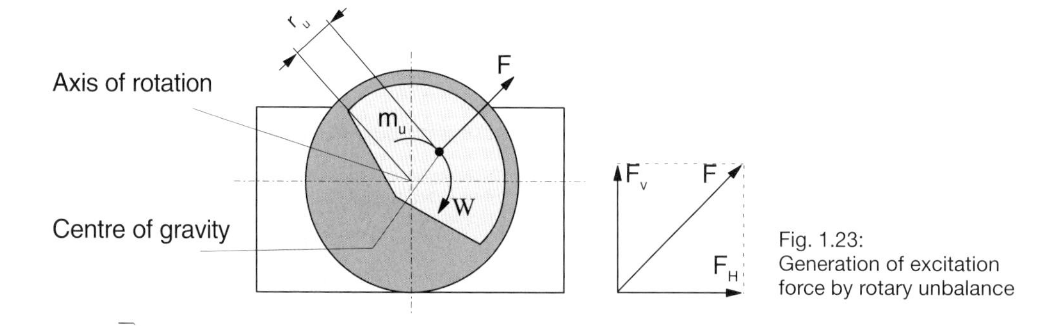

Fig. 1.23:
Generation of excitation force by rotary unbalance

and the calibration ratio,

$$\eta = \frac{\Omega}{\omega} \tag{1.14}$$

gives the following for the stationary solution:

$$z = \frac{\hat{F}}{c} \frac{1}{\sqrt{(1-\eta^2)^2 + 4D^2\eta^2}} \sin(\Omega t - \varphi) \tag{1.15}$$

where D = damping

For the assessment of concrete product compaction, for instance, it is initially sufficient to consider the amplitude of movement $\hat{z}$ of the mass m, which means that the phase shift φ between the excitation force and the motion is not considered at this stage. Application of Equation (1.15) to Equations (1.10), (1.13) and (1.14) gives:

$$\hat{z} = \frac{m_u \cdot r_u}{m} \frac{\eta^2}{\sqrt{(1-\eta^2)^2 + 4D^2\eta^2}} = \frac{m_u \cdot r_u}{m} V_1 \tag{1.16}$$

where V_1 is the magnification factor for the motion of the mass m.

In addition to the motion of the mass m, another parameter required to set up the equipment is the dynamic force F_B transferred to the place of installation, which results from Fig. 1.22 as follows:

$$F_B = cz + k\dot{z}. \tag{1.17}$$

This gives the amplitude of the force acting on the floor as follows:

$$\hat{F}_B = \frac{m_u \cdot r_u}{m} c\eta^2 \sqrt{\frac{1 + 4D^2\eta^2}{(1-\eta^2)^2 + 4D^2\eta^2}} = \frac{m_u \cdot r_u}{m} cV_2 \tag{1.18}$$

where V_2 is the magnification factor for the force F_B acting on the floor.

Both the equipment oscillation and the force acting on the point of installation are thus determined dynamically using the magnification factors V_1 and V_2. Fig. 1.24 shows each of these factors as a function of the calibration ratio η.

The following paragraphs discuss Equations (1.16) and (1.18).

As mentioned above, damping may be neglected for the purpose of outlining underlying principles and relationships. For Equation (1.16), this gives:

$$V_1 = \frac{\hat{z} \cdot m}{m_u \cdot r_u} = \frac{\eta^2}{1-\eta^2} \tag{1.19}$$

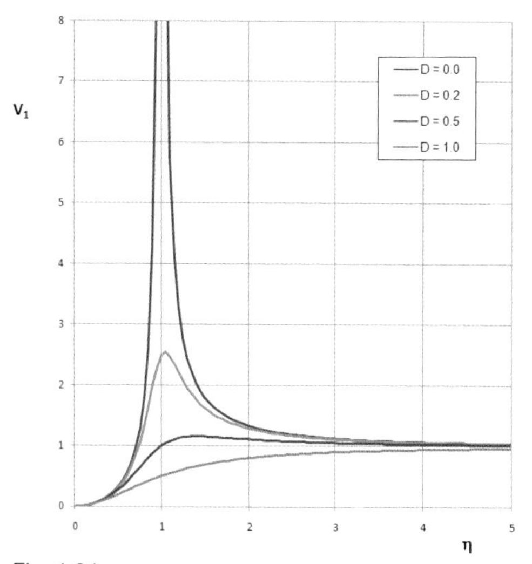

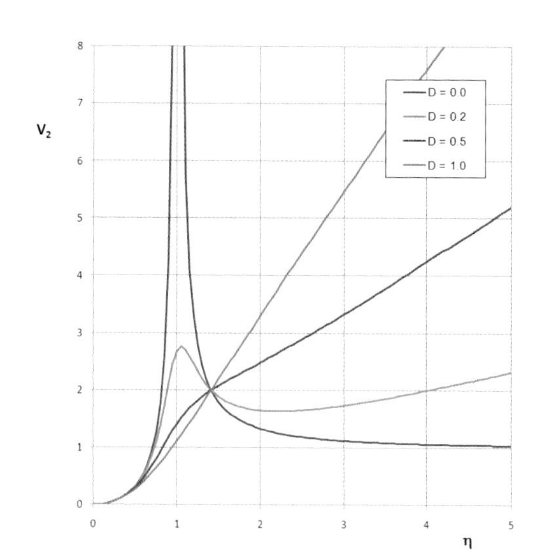

Fig. 1.24:
Magnification factors V_1 and V_2
V_1 magnification factor for the motion of the mass m
V_2 magnification factor for the force F_B acting on the floor

and for the amplitude of displacement:

$$\hat{z} = \frac{m_u \cdot r_u}{m} \cdot \frac{\eta^2}{1-\eta^2} \qquad (1.20)$$

Equation (1.20) can be used to calculate the amplitude of vibration acceleration â. This parameter serves to determine characteristic values for the evaluation of vibration systems and to compare them with recommended values.

$$\hat{a} = \hat{z} \cdot \Omega^2 = \frac{m_u \cdot r_u}{m} \cdot \frac{\eta^2}{1-\eta^2} \cdot \Omega^2 \qquad (1.21)$$

Equation (1.21) thus also provides the parameters required to achieve specified motion characteristics.

The following two options exist to achieve certain amplitudes of movement $\hat{z}$ and $\hat{a}$ because the total mass m is usually defined by the equipment design, and the angular excitation frequency Ω results from the specific processing parameters:

- selection of a corresponding excitation force F by varying the unbalance mass m_u and the unbalance radius r_u using Equation (1.10), or
- selection of a suitable calibration ratio η.

Because the selection of η << 1 (i.e. a sub-critical mode of operation, or a system calibrated to high frequencies) results in V_1 values that are too low, the only remaining option is to operate the system in the supercritical range at η >> 1 (i.e. a system calibrated to low frequencies). A range from 3 to 5 is usually specified for the calibration ratio η In this case, we can assume that $V_1 \approx 1$, as can be concluded from Equation (1.16) and Fig. 1.24.

Equation (1.20) then results in the centre-of-mass theorem:

$$\hat{z}m = m_u \cdot r_u \qquad (1.22)$$

Taking account of Equation (1.10), Equation (1.21) can be used to derive

$$\hat{a}m = m_u \cdot r_u \cdot \Omega^2 = \hat{F} . \qquad (1.23)$$

Although the relevant literature mostly cites this equation without further comment, it only applies when $\eta >> 1$ and $D = 0$.

Operation at $\eta > 3$ is also favourable with regard to the dynamic forces F_B that act on the place of installation in accordance with Fig. 1.22 and Equation (1.18). However, the strong damping effect in this area must be taken into account. Given that

$$\hat{\dot{z}} = \Omega\hat{z} \qquad (1.24)$$

the damping force F_D and thus the dynamic force F_B increase with the rising angular excitation frequency Ω or the increase in η in accordance with Equation (1.17). This means that low damping levels should be achieved in order to maintain low forces F_B acting on the environment in the supercritical range.

1.1.5 Process Parameters

There is a very large number of production process parameters. This set of parameters depends on the objects, on the type of the specific production process and on the purpose of its use. For this reason, the following section concentrates on some of the key parameters.

1.1.5.1 Parameters determining the process macrostructure

Spatial parameters

- production area A_p, area → utilised by the production process
- specific production area $A_{Pspez} = \frac{A_p}{M}$,area → utilised for the production of one quantity unit of the product

Temporal parameters

- mould cycle time t_u
- handling time of the consolidation unit t_{uv}
- cycle time t_i

Quantity parameters

- produced quantity

$$M = \sum_{i=1}^{n} = M_i$$

Total quantity of products resulting from the manufacturing process with the subsets M_i of product types i

- production output

$$P = \dot{M}$$

$\dot{M} = \frac{dM}{dt}$ quantity flow or volumetric flow $\dot{V} = \frac{dV}{dt}$ in m^3/h and mass flow

$\dot{m} = \frac{dm}{dt}$ in $\frac{t}{h}$

- number of moulds
  number of moulds in the production process

- number of bays
  - o number of bays in the moulding process
  - o number of bays in the consolidation process

1.1.5.2 Parameters determining the process microstructure

These include parameters for vibratory compaction and consolidation. The following section deals with the parameters for vibratory compaction.

a) Parameters for vibratory compaction of the concrete mix

Measurable parameters that enable correlations with the specified fresh and hardened concrete properties are required to assess the vibrational pattern during vibratory compaction. These parameters have been defined on the basis of harmonic oscillation. A closer look at them reveals that they always include the motion parameters, in particular acceleration values, and the frequencies of the individual oscillations. They thus constitute primary parameters.

The following vibratory compaction parameters are commonly applied today:

1. Frequency of oscillation
   - excitation frequency f
   - angular excitation frequency $\Omega = 2 \cdot \pi \cdot f$

2. Motion parameters
   - vibration displacement z
     amplitude of vibration displacement $\hat{z}$
   - vibration velocity v
     amplitude of vibration velocity $\hat{v}$
   - vibration acceleration a
     amplitude of vibration acceleration $\hat{a}$,
     often expressed as the relative acceleration

     $$a_g = \frac{\hat{a}}{g}$$

     where g: gravitational acceleration

3. Compaction time t_v

4. Compaction intensity

$$I_v = \hat{z}^2 \cdot f^3 \tag{1.25}$$

5. Overall compaction

$$W_V = I_V \cdot t_V \tag{1.26}$$

A certain compaction intensity I_v can be achieved by various combinations of frequencies and amplitudes of displacement. By the same token, a desired overall compaction W_v can be achieved by totalling the various combinations of intensities and compaction times. Modern methods use this approach to take account of the individual compaction phases. Today, many types of vibration equipment enable the use of this method because they provide options to control frequencies and to continuously adjust the excitation force, for instance, by forced synchronisation of the phase positions of several vibrators using an electronic system.

Relevant rules and standards usually specify frequency-dependent acceleration values. For example, DIN 4235 Part 3:1978-12 [1.9], which governs the moulding and compaction of precast elements subjected to in-mould curing, gives standard values of the acceleration amplitude on the mould surface (area of contact with the concrete) as a function of the excitation frequency (Table 1.3).

Table 1.3: Standard acceleration values in accordance with DIN 4235

Excitation frequency f [Hz]	Acceleration â in m/s²
50	30 to 50
100	60 to 80
150	80 to 100
200	100 to 120

These parameters do not apply, however, to block and pipe machines that process zero-slump concrete mixes and include a demoulding step while the concrete is still fresh.

Problems arise with standard values if pulsed excitation (shock vibration) is used. Related parameters and standard values have not yet been defined.

The vibration-induced motion behaviour within the concrete mix must be considered when selecting appropriate excitation frequencies. Chapter 1.4 outlines the two basic approaches applied in this regard, i.e. the phenomenological and corpuscular models.

Additional parameters are required to control the compaction process as well as the resulting design and engineering of the compaction equipment to be used. The parameters that determine the moulding and compaction of precast products are grouped into classes.

b) Parameter classes

The parameter classes for vibratory compaction are shown in Figs. 1.24, 1.25 and 1.26. The classes referred to in these figures form a causal chain.

Equipment parameters include masses, stiffnesses, damping coefficients and forces. These vibration system parameters determine the motion behaviour of the working masses, which is characterised by variables such as acceleration amplitudes, types of motion or phase positions of the moving pipe core and jacket.

The motions of the working masses define the parameters for the actions on the concrete. In the concrete mix, the actions on its edge determine the physical parameters that act as internal compaction parameters to trigger compaction of the entire mix.

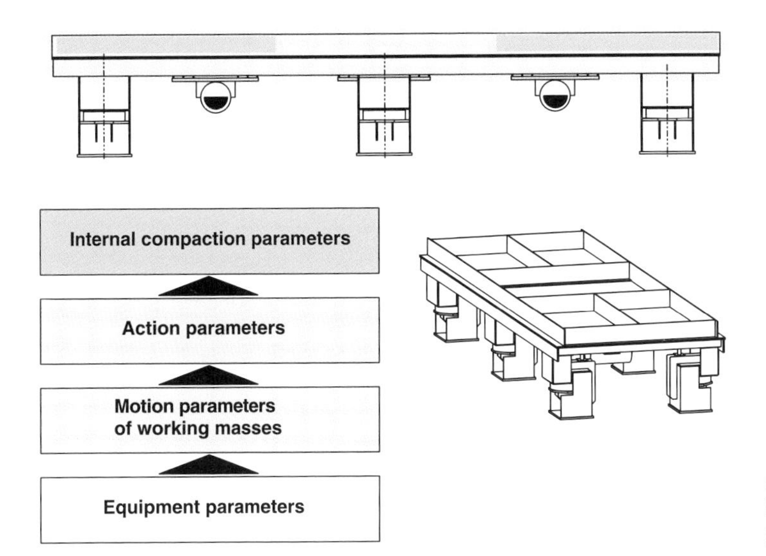

Fig. 1.25:
Parameter classes in precast element production

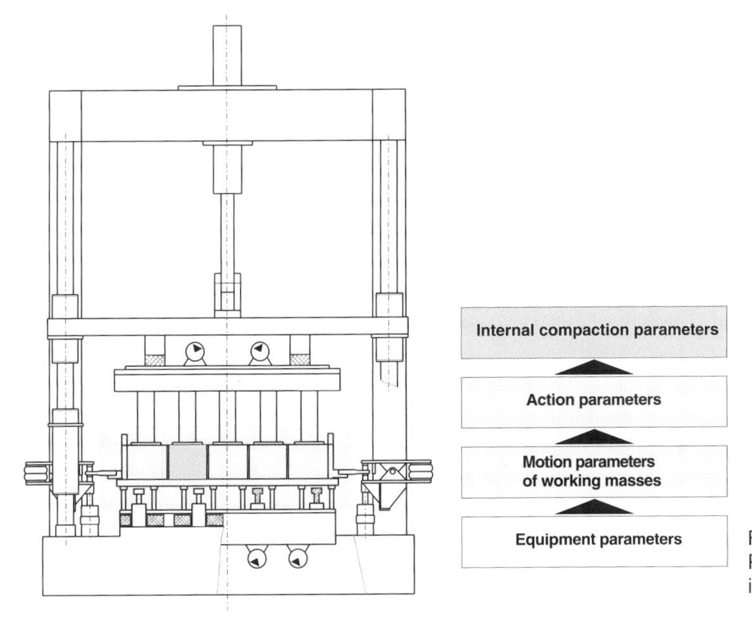

Fig. 1.26:
Parameter classes in block production

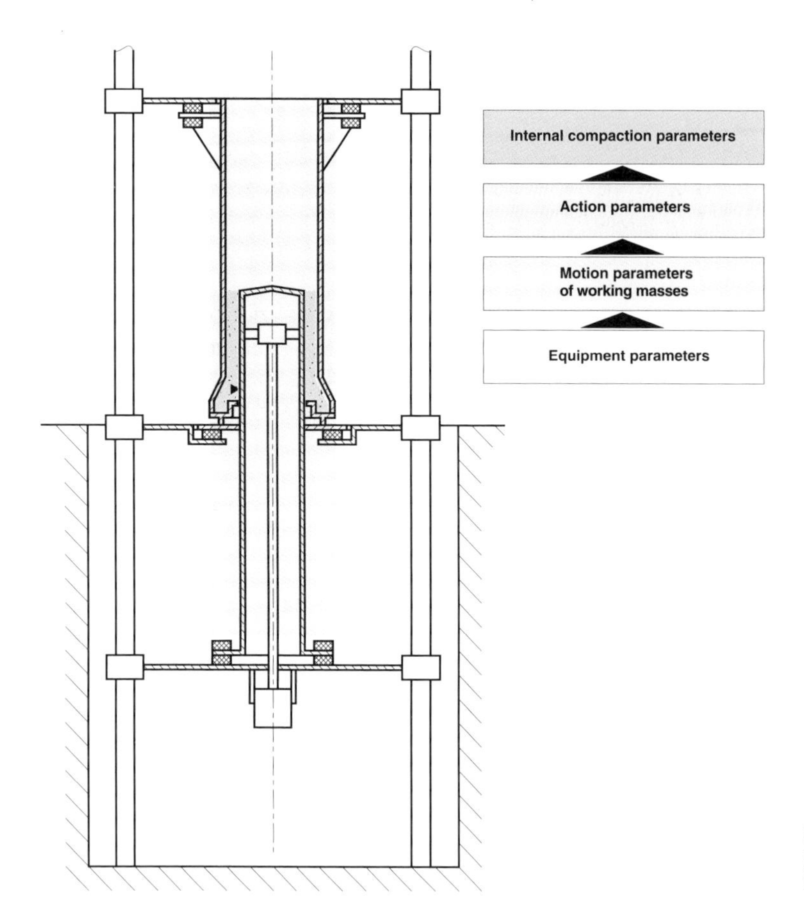

Fig. 1.27:
Parameter classes
in pipe production

Internal compaction parameters
Internal compaction parameters are thus physical parameters that cause compaction of a concrete unit. This unit is located within the mix. Only the physical parameters that occur on the element edges are relevant to the compaction of the precast unit. These parameters result from external forces and their subsequent transmission within the concrete. For instance, internal compaction parameters can be used to explain differences in compaction within a single structural element.

Action parameters
Action parameters are physical parameters that act on the surface zone of the concrete mix, i.e. at the interfaces between the concrete and the equipment. In the case of pipe machines, these include the interfaces between the core and the concrete, between the jacket and the concrete or between the top and bottom ring and the concrete. Action parameters include motion and stress values analogous to the internal compaction parameters. Just like these, they must be considered as a function of time. This approach results in additional variables, such as energy input and output.

Motion parameters of working masses
Working masses are the parts and components of the vibration equipment that actually vibrate. In pipe machines, these usually include the core, the jacket and the top and bottom ring. In block machines, the working masses include the table, board, mould and tamper head. The motions of the working masses are described by their magnitude, progress over time and inherent frequency.

Equipment parameters
Equipment parameters include all parameters that influence the motion behaviour of the working masses, such as masses, stiffnesses, damping coefficients, bracing and exciter forces and impact intervals. This category also includes the properties of the concrete mix that are relevant to the motion of the working masses because the concrete mix is also a component of the vibration system.

1.2 Fundamentals of Materials

1.2.1 Raw Materials for the Production of the Concrete Mix

Concrete is the most frequently used construction material and can be considered an artificial stone. In its basic design, it consists of the main constituents cement, aggregate and water. State-of-the-art concrete grades are quinary systems that also include additives and admixtures. These ingredients ensure compliance with specific requirements such as concrete workability and the characteristics of the hardened concrete.

1.2.1.1 Cement

Cement is a hydraulic binder. It hardens after mixing with water, both when exposed to air and underwater.

Cement is produced by sintering finely ground raw materials (limestone, clay, silica sand, marl, iron ore) and milling the resulting Portland cement clinker. It is essentially composed of the four clinker phases:

- tricalcium silicate (alite) C_3S
- dicalcium silicate (belite) C_2S
- tricalcium aluminate C_3A
- calcium aluminate ferrite C_4AF

Regional standards specify the composition of the cement. DIN EN 197-1:2004-08 [1.10] applies to Europe. DIN 1045-2:2008-08 [1.17] specifies the areas of application for standard cement used for the production of concrete. Table 1.4 provides an overview of the cement grades covered by DIN EN 197-1.

In addition, DIN 1164 specifies cements with special characteristics. These are:

- highly sulphate-resistant cements (HS cements in accordance with DIN 1164-10:2004-08) [1.11],
- cements with a low effective alkali content (NA cements in accordance with DIN 1164-10:2004-08),
- cements with a low heat of hydration (NW cements in accordance with DIN 1164-10:2004-08),
- quick-set cements (FE and SE cements in accordance with DIN 1164-11: 2003-11) [1.12],
- cements with an increased ratio of organic constituents (HO cements in accordance with DIN 1164-12:2005-06) [1.13].

Table 1.4: Cement grades and their composition in accordance with DIN EN 197-1: 2004-08 [1.10]

Cement grade			Main constituents other than Portland cement clinker (K)	
Main grade	Name	Abbreviated designation	Type	Proportion in M.-%
CEM I	Portland cement	CEM I	-	0
CEM II	Portland slag cement	CEM II/A-S	blast-furnace slag (S)	6…20
		CEM II/B-S		21…35
	Portland silica fume cement	CEM II/A-D	silica fume (D)	6…10
	Portland pozzolanic cement	CEM II/A-P	natural pozzolana (P)	6…20
		CEM II/B-P		21…35
		CEM II/A-Q	artificial pozzolanic material (Q)	6…20
		CEM II/B-Q		21…35
	Portland fly ash cement	CEM II/A-V	siliceous fly ash (V)	6…20
		CEM II/B-V		21…35
		CEM II/A-W	calcareous fly ash (W)	6…20
		CEM II/B-W		21…35
	Portland shale cement	CEM II/A-T	burnt shale (T)	6…20
		CEM II/B-T		21…35
	Portland limestone cement	CEM II/A-L	limestone (L)	6…20
		CEM II/B-L		21…35
		CEM II/A-LL	limestone (LL)	6…20
		CEM II/B-LL		21…35
	Portland composite cement	CEM II/A-M	all main constituents possible (S, D, P, Q, V, W, T, L, LL)	6…20
		CEM II/B-M		21…35
CEM III	Blast-furnace cement	CEM III/A	blast-furnace slag (S)	36…65
		CEM III/B		66…80
		CEM III/C		81…95
CEM IV	Pozzolanic cement	CEM IV/A	pozzolanic materials (D, P, Q, V,) silica fume (D), fly ash (V, W) possible	11…35
		CEM IV/B		36…55
CEM V	Composite cement	CEM V/A	blast-furnace slag (S) and pozzolanic materials (P, Q, V) possible, including siliceous fly ash (V)	18…30
		CEM V/B		31…50

Table 1.5: Cement strength classes in accordance with DIN EN 197-1:2004-08 [1.10]

Strength class	Compressive strength [N/mm²]			
	Initial strength		Standard strength	
	2 days	7 days	28 days	
32.5 N	-	≥ 10	≥ 32.5	≤ 52.5
32.5 R	≥ 10	-		
42.5 N	≥ 10	-	≥ 42.5	≤ 62.5
42.5 R	≥ 20	-		
52.5 N	≥ 20	-	≥ 52.5	-
52.5 R	≥ 30			

The standardised cement designation must always include the cement grade, the referenced standard, the abbreviated designation of the cement grade, the strength class (as shown in Table 1.5) and, if applicable, any special characteristics, e.g. Portland cement DIN 1164 CEM I 42,5 R-HS.

Other important properties of the cement include its setting behaviour, volume stability, heat of hydration, colour, density and bulk density, and particle fineness.

1.2.1.2 Aggregates

The term aggregate is used to refer to a granular material to be used in the construction industry. Aggregates can be distinguished with respect to their origin, bulk density and particle size (see Table 1.6).

Table 1.6: Classification of aggregates

Classification according to	Aggregate	Definition/specification
Origin	natural	naturally occurring mineral, mechanical processing only
	industrially produced	mineral origin, industrially produced (thermal or other process)
	recycled	inorganic material processed from construction waste, generic term to refer to recycled chippings and recycled crushed sand
	gravel	naturally rounded material
	chippings	crushed material
Bulk density	normal	particle bulk density > 2,000 kg/m³ mineral origin
	lightweight	particle bulk density ≤ 2,000 kg/m³ or bulk density ≤ 1,200 kg/m³ mineral origin
Particle size	coarse	D ≥ 4 mm and d ≥ 2 mm
	fine	D ≤ 4 mm (sand)
	fines ratio	rock ratio < 0.063 mm
	filler (rock powder)	major fraction < 0.063 mm

Aggregates are also categorised according to their particle size ranges (or product sizes). Size ranges are stated according to defined sets of basic screens or sets of basic and supplementary screens.

Other limit values are applied to aggregates in the following categories:

- particle composition (particle size distribution indicated by undersizes)
- threshold deviations from the typical particle composition
- fines ratio
- freeze-thaw resistance
- magnesium sulphate resistance
- ratio of lightweight organic contaminations
- flakiness index
- particle shape index
- acid-soluble sulphate content
- mussel shell content in coarse aggregates

(See e.g. Table 1.7.) DIN EN 12620:2008-07 [1.14] applies.

DIN EN 13055-1:2002-08 [1.15] defines special requirements for lightweight aggregates, whereas DIN 4226-100:2002-02 [1.16] includes requirements for recycled aggregates.

Specific concrete technology parameters are defined in addition to the general aggregate specifications. These parameters include a defined grading curve (Fig. 1.28) but also the so-called fineness modulus or grading coefficient (k value) and cumulative fraction that has passed through a particular mesh (D total), as well as the water demand.

Table 1.7: Classification of freeze-thaw resistance in accordance with DIN EN 12620:2008-07 [1.14]

F category	Mass loss [M.-%][1)]
F_1	≤ 1
F_2	≤ 2
F_4	≥ 4
$F_{specified}$	> 4
F_{NR}	no requirements

[1)] In extreme situations, the test specified in DIN EN 1367-1:1999, Annex B, may be carried out using a saline solution or urea. However, the threshold values stated in this table do not apply to such cases.

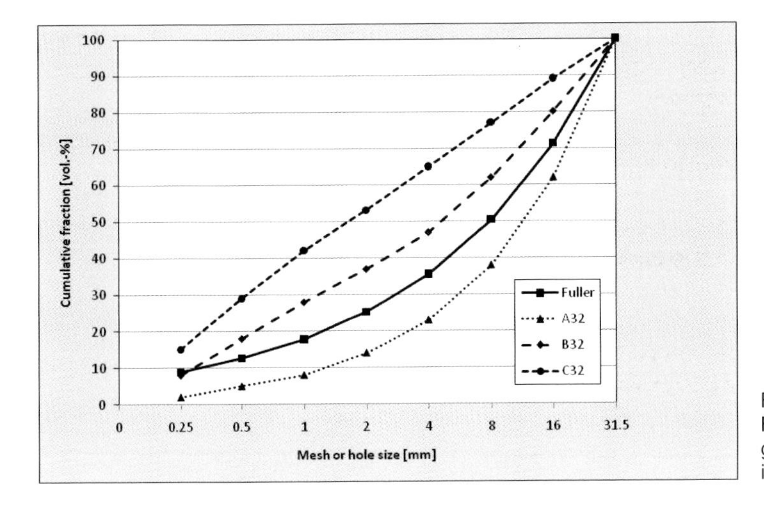

Fig. 1.28:
Fuller curve and standard grading curves as specified in DIN 1045:2008-08 [1.17]

1.2.1.3 Concrete admixtures

Concrete admixtures are finely dispersed materials that are used in concrete to achieve or improve certain characteristics. DIN EN 206-1:2001-07 [1.18] distinguishes two types of inorganic admixtures:

Type I: practically inert admixtures, such as rock powders in accordance with DIN EN 12620:2008-07 [1.14] or pigments in accordance with DIN EN 12878:2006-05 [1.19]

Type II: pozzolanic or latent hydraulic admixtures, such as trass (see Table 1.8) according to DIN 51043:1979-08 [1.20], fly ash (Table 1.9) according to DIN EN 450-1: 2008-05 [1.21] or silica fume (Table 1.9) according to DIN EN 13263-1:2009-07 [1.22]

Table 1.8: Technical parameters of trass and limestone powder

Technical parameters	Unit	Trass (DIN 51043)	Limestone powder
Specific surface area	cm^2/g	≥ 5,000	≥ 3,500
Particle fraction < 0.063 mm	M.-%	-	≥ 70
Loss on ignition	M.-%	≤ 12	~ 40
Sulphate (SO_3)	M.-%	≤ 1.0	≤ 0.8
Chloride (Cl^-)	M.-%	≤ 0.10	≤ 0.04
Density [1)]	kg/dm^3	2.4 … 2.6	2.6 … 2.7
Bulk density [1)]	kg/dm^3	0.7 … 1.0	1.0 … 1.3

[1)] DIN 1045-2: 2008-08 only permits ignition loss category A (≤ 5.0 M.-%)

Table 1.9: Technical parameters of fly ash and silica fume

Technical parameters	Unit	Fly ash	Silica fume	
			Powder	Suspension
Fineness (> 0.045 mm) Category N: Category S:	M.-%	≤ 40 ≤ 12	-	-
Specific surface area	cm^2/g	-	≥ 150,000 ≤ 350,000	-
Loss on ignition	M.-%	≤ 5.0 [1)]	≤ 4.0	≤ 4.0
Sulphate (SO_3)	M.-%	≤ 3.0	≤ 2.0	≤ 2.0
Chloride (Cl^-)	M.-%	≤ 0.10	≤ 0.30 [2)]	≤ 0.30 [2)]
Alkaline constituents (Na_2O equivalent)	M.-%	≤ 5.0	manufacturer's specification	manufacturer's specification
Density [3)]	kg/dm^3	2.2 … 2.6	ca. 2.2	ca. 1.4
Bulk density [3)]	kg/dm^3	1.0 … 1.1	0.3 … 0.6	-

[1)] DIN 1045-2:2008-08 only permits ignition loss category A (≤ 5.0 M.-%)
[2)] Cl^- ratios in excess of 0.10 M.-% must be declared; in the case of Cl^- ratios greater than 0.20 M.-%, compliance with DIN 1045-2:2008-08, Table 1.9, is required for concrete with prestressing steel
[3)] Reference values from experience gained to date

Further admixtures include plastic dispersions and fibres (Table 1.10 and Table 1.11).

Table 1.10: Classification and characteristics of steel fibres in accordance with DIN EN 14889-1:2006-11 [1.23]

Steel fibres according to DIN EN 14889-1		
Classification according to manufacturing method	Group I	cold-drawn steel wire
	Group II	fibres cut from sheet steel
	Group III	fibres extracted from molten material
	Group IV	fibres cut from cold-drawn wire
	Group V	fibres shaved from steel ingots
Description using the following characteristics	group and shape	
	geometry: length and equivalent diameter	
	tensile strength and modulus of elasticity	
	ductility (if required)	
	influence on concrete workability (reference concrete)	
	influence on tensile bending strength (reference concrete)	

Table 1.11: Classification and characteristics of polymer fibres in accordance with DIN EN 14889-2:2006-11 [1.24]

Polymer fibres according to DIN EN 14889-2		
Classification according to physical shape	Class Ia	microfibres with d < 0.30 mm (monofilament fibres)
	Class Ib	microfibres with d < 0.30 mm (fibrillated fibres)
	Class II	macrofibres with d > 0.30 mm
Description using the following characteristics	class, type of polymer, shape, bundling and surface finish	
	geometry, length, equivalent diameter and fineness (Class I)	
	force relative to fineness (Class I) / tensile strength (Class II), modulus of elasticity	
	melting point and flash point	
	influence on concrete workability (reference concrete)	
	influence on tensile bending strength (reference concrete)	

1.2.1.4 Concrete additives

Concrete additives are powders or liquids that are added to the concrete in small quantities during the mixing process. They modify the chemical and physical properties of the fresh and/or hardened concrete. During the past few years in particular, many novel, more advanced additives have been launched. Without them, it would have been impossible to develop easily compactable concrete (ECC) and self-compacting concrete (SCC), but also high-performance and ultra-high performance concrete (UHPC). Table 1.12 lists the range of possible concrete additives.

Table 1.12: Types of concrete additives classified according to their mechanism of action

Type/mechanism of action	Abbreviated designation
Concrete workability agents	CWA
Plasticisers	P
Air-entraining agents	AEA
Waterproofing agents	WPA
Retarders	R
Setting/hardening accelerators	S/HA
Shotcrete setting accelerators	SSA
Grouting aids	GA
Stabilisers	ST
Sedimentation reducers	SR
Chromate reducers	CR
Foaming agents	FA
Elastic hollow spheres for air-entrained concrete	
Expansion aids	
Sealants	
Passivating agents	

1.2.1.5 Mixing water

According to DIN EN 1008:2002-10 [1.25], the following types of mixing water are suitable for concrete:

- drinking water
- groundwater
- natural surface water
- industrial water
- residual water from recycling plants in concrete production
- seawater or brackish water (only for non-reinforced concrete)

The mixing water must be tested for suitability, except in the case of drinking water. During the mixing water tests, not only the preliminary testing requirements (Table 1.13) but also chemical specifications (Table 1.14) and defined setting time and compressive strength parameters (Table 1.15) must be adhered to.

Table 1.13: Requirements for the preliminary testing of mixing water

Criterion	Requirements
Oil and grease	only traces
Detergents	foam must collapse within a period of two minutes
Colour	clear to slightly yellowish (except residual water)
Suspended solids	residual water as specified in DIN EN 1008, Table 5.2.a other water: ≤ 4 ml of settling volume
Odour	residual water: only drinking water odour and slight cement odour, or slight hydrogen sulphide odour if water contains fly ash other water: only drinking water odour: no hydrogen sulphide odour after addition of hydrochloric acid
Acids	pH ≈ 4
Humic matter	colour not more than slightly yellowish-brown after addition of NaOH

Table 1.14: Chemical specifications for mixing water

Chemical characteristic	Maximum content [mg/l]
Chloride (Cl^-) Prestressed concrete/grouting mortar Reinforced concrete Non-reinforced concrete	 ≤ 500 ≤ 1,000 ≤ 4,500
Sulphates (SO_4^{2-})	≤ 2,000
Na_2O equivalent	≤ 1,500
Contaminants with a deleterious effect on concrete: Sugar Phosphates (P_2O_5) Nitrates (NO_3^-) Lead (Pb^{2+}) Zinc (Zn^{2+})	 ≤ 100 ≤ 100 ≤ 500 ≤ 100 ≤ 100

Table 1.15: Setting time and compressive strength requirements for the testing of mixing water

Criterion	Requirement
Setting times	Start of setting ≥ 1 hour End of setting ≤ 12 hours Deviation ≤ 25% from the test value obtained with distilled or deionised water
Mean compressive strength after 7 days	≥ 90% of mean compressive strength of test specimens with distilled or deionised water

1.2.2 Concrete Mix Design and Composition

a) Basics

According to generally accepted terminology, concrete is termed fresh as long as it is still workable and compactable. Since the authors of this book are focussing on processes occurring during placement and compaction of concrete mixes for the production of premium concrete products, the following additional definitions are introduced:

The term "concrete mix" refers to the mix from the time of mixing to its placement in the mould or formwork. The term fresh concrete refers exclusively to fully compacted concrete.

As described in Section 1.2.1, concrete has evolved from a ternary to a quinary system. Thus the options to vary the characteristics of the concrete mix and the properties of fresh and hardened concrete are virtually unlimited.

DIN EN 206-1:2001-07 [1.18] states that the concrete composition and raw materials must be chosen so as to fulfil the requirements defined for concrete works with respect to both the concrete mix and the fresh and hardened concrete whilst taking account of the production process and selected execution method. These requirements include workability, bulk density, strength, durability and protection of the embedded steel against corrosion.

b) Calculation of the material volume

Compressive strength and bulk density are the characteristics that determine the classification of the individual concrete grades. They are also the key parameters that influence the design of the concrete, and are determined using a material volume calculation.

This method forms the basis for the mix design and is used to calculate the composition of the fresh concrete volume. One cubic metre of compacted fresh concrete is taken as the reference. The volume of raw materials is used in the calculation, whereas the moisture of the aggregates is not taken into account. The following equation is used for the calculation; it expresses the functional relationship between the volumes and masses of the multi-component system:

$$1000 = \frac{z}{\rho_z} + \frac{f}{\rho_f} + \frac{w}{\rho_w} + \frac{g}{\rho_g} + p \left[\frac{dm^3}{m^3}\right] \quad (1.27)$$

where:

z	cement content	[kg/m³]
f	additives content (e.g. fly ash, silica fume, rock powder)	[kg/m³]
w	water content	[kg/m³]
g	aggregate content	[kg/m³]
p	pore volume	[dm³/m³]
ρ_z	density of cement	[kg/dm³]
ρ_f	density of additives	[kg/dm³]
ρ_w	density of water	[kg/dm³]
ρ_g	bulk density of aggregates	[kg/dm³]

The starting point for a mix design is the target compressive strength of the concrete product. The compressive strength of concrete is influenced by the water/cement ratio and the standard compressive strength of the cement. The relationship between these two factors is given by so-called Walz curves. These curves indicate the water/cement ratio required to achieve a specific compressive strength for a given standard compressive strength of the cement used in the mix. The influence of the standard compressive strength of the cement becomes less significant in high-strength concretes. The target compressive strength should be selected so as to ensure that the specified minimum and/or maximum values of the relevant performance characteristics are adhered to with a sufficient degree of certainty.

The next step of the calculation determines the required amount of water, which depends on:

- particle composition (grading curve)
- maximum particle size (the coarser the aggregate mix and the larger the maximum particle size, the lower the water demand)
- particle shape and surface
- powder ratio
- concrete additives and admixtures used
- specified workability.

The aggregate grading curve can be used to determine the cumulative fraction D (total of all sizes passing through the screen) and the grading coefficient k (total of the percentage residues on a set of screens with the sizes 0.25 - 0.5 - 1 - 2 - 4 - 8 - 16 - 31.5 - 63 mm, divided by 100). Both parameters are necessary to subsequently calculate the water demand of the mineral aggregate whilst considering the specified workability class. Reference [1.28] specifies standard values for the water demand of the concrete mix as a function of the particle composition and concrete workability.

In the third step, the quantity of cement required for the ternary system is calculated using:

$$z = \frac{w}{\omega} \tag{1.28}$$

For the quinary system, we use

$$\omega_{eq} = \frac{w}{z + k_f \cdot f + k_s \cdot s} \tag{1.29}$$

$$z = \frac{w}{\omega_{eq}} - k_f \cdot f - k_s \cdot s \tag{1.30}$$

where:

z	cement content	[kg/m³]
w	water content	[kg/m³]
ω	water/cement ratio	[-]
ω_{eq}	equivalent water/cement ratio	[-]
f	fly ash content	[kg/m³]
s	silica fume content	[kg/m³]
k_f	efficiency factor for fly ash	
k_s	efficiency factor for silica fume	

If concrete additives and admixtures are added, account must be taken of the permissible maximum quantities (i.e. quantities that may be considered for the purpose of the calculation). The following maximum quantities may be considered for fly ash and silica fume:

Fly ash: $max_{fb} \leq 0.33 \cdot z$
Silica fume: $max_{sb} \leq 0.11 \cdot z$

The maximum quantities are determined by the powder content. When adding liquid admixtures, the water ratio must be considered in the material volume calculation if the admixture ratio is greater than 3.0 l/m³.

In the last step, the material volume calculation is used to determine the required amount of aggregates. The individual proportions can be quantified by applying the rule of mixture and the grading coefficients of the individual aggregate sizes. This forms the basis for calculating the required bulk density of the fresh concrete.

At this stage, the mix design may still be inappropriate owing to variance in the water demand of the individual aggregates and can be evaluated only after initial testing.

c) Concrete mix design according to prior specification
DIN EN 206-1:2001-07 [1.18] and DIN 1045-2:2008-08 [1.17] state that the author of the specification determines the requirements for the concrete to be designed. These requirements include not only its strength and workability class, but also the exposure classes, terms of use and details of the maximum aggregate size and the type of use. A distinction is made between:

- standard concrete
- concrete determined by composition
- concrete determined by characteristics

Standard concrete is a standardised, low-strength concrete (up to C16/20) with additional specifications (X0, XC1, XC2). No additives or admixtures may be used. The specified minimum cement quantities must be adhered to. Only natural aggregates may be used. No tests of the concrete mix, fresh and hardened concrete are required during placement on the construction site.

Concrete grades determined by their composition may extend to all strength and exposure classes. The author of the specifications (i.e. the client or specifier) determines the composition of the concrete and is thus responsible for ensuring that the specified concrete fulfils all strength and durability requirements. He/she is also responsible for initial testing. The producer must verify compliance of the mix with the relevant standard. Concrete determined by composition should only be used in special buildings or structures after it has been subjected to comprehensive concrete engineering tests.

Concrete determined by characteristics may also extend to all strength and exposure classes. The concrete producer is responsible for initial testing and thus also specifies the necessary characteristics and additional requirements whilst also verifying conformity. The major share of the concrete placed in Germany falls under this category.

During its life cycle, concrete is subjected to varying ambient conditions. These ambient and corrosion conditions comprise physical, chemical and mechanical impacts. They act on the concrete and its reinforcement and cannot be captured by design loads. These influences were classified and grouped into so-called exposure classes in DIN EN 206-1:2001-07 [1.18] and DIN 1045-2:2008-08 [1.17]. The following exposure classes exist:

- X0 no risk of corrosion and attack
- XC reinforcement corrosion triggered by carbonation

- XD reinforcement corrosion triggered by chlorides (except seawater)
- XS reinforcement corrosion triggered by chlorides in seawater
- XF concrete corrosion caused by frost attack with or without de-icing agent
- XA concrete corrosion by chemical attack
- XM concrete erosion by wear and tear

For the purpose of concrete design, these exposure classes are supported by defined limit values, which include, in particular, the maximum water/cement ratio, minimum cement quantities, air void ratios and compressive strength parameters.

In addition, the concrete design must consider the anticipated environmental conditions by assigning moisture classes to the concrete in order to take preventative measures against any damage that may be caused by alkali-silica reactions. DIN 1045-2:2008-08 [1.17] specifies four moisture classes.

The water/cement ratio is determined using the maximum permissible water/cement ratios defined in the exposure classes as well as the Walz curve value derived from the target compressive strength and the standard compressive strength of the cement. The concrete design must always be based on the lower of these two values.

The further procedure is identical to that described in Section 1.2.2 – the material volumes are calculated. The minimum cement contents stated for the exposure classes must also be considered when determining the required amount of cement.

The durability of the concrete for the intended use under the prevailing local conditions is deemed verified if both concrete composition and compressive strength class comply with the specifications. The following preconditions apply:

- the appropriate exposure and moisture classes have been selected
- the concrete has been poured and compacted in accordance with applicable rules and standards
- the minimum concrete cover has been adhered to
- appropriate maintenance measures have been taken.

1.2.3 Concrete Properties

1.2.3.1 Properties of the concrete mix/fresh concrete

a) Consistency and workability

The sub-processes of mixing, transport, moulding and compaction determine the key requirements for the concrete mix and fresh concrete.

Workability is not defined in physical terms and cannot be measured directly. This notion includes both the rheological properties (i.e. viscosity, yield limit, internal friction) and the behaviour of the concrete mix during mixing, transport, moulding and compaction. Good workability is assumed if the concrete shows good cohesion and no segregation phenomena, and is fully compactable. Workability must be tailored to the specific application [1.29].

As a quantitative measure describing the workability and working time of the concrete, consistency is considered an important characteristic of the concrete mix to ensure trouble-free transport, spreading and compaction of the mix and to achieve a good surface finish. The main factors that determine the consistency of the concrete mix are its rheological properties and the volume of the cement paste, as well as the type and particle composition of the aggregate. In practice, this adjustment is initially made by defining the water/cement ratio, which expresses the mass ratio between the effective water content and the cement content in relation to one cubic metre of compacted fresh concrete [1.30].

However, the water/cement ratio alone is not sufficient to determine the consistency and workability of the concrete mix with reasonable certainty. The addition of water increases not only the water/cement ratio, but also the amount of cement paste, which has an additional influence on the consistency of the mix.

Key factors that influence the consistency include:

- water content
- cement content
- amount of cement paste
- water demand of the concrete raw materials

Table 1.16: Consistency ranges and classes in accordance with DIN 1045-2:2008-08 [1.17]

Consistency description	Slump		Compacting factor	
	Class	Slump (diameter)	Class	Compacting factor
	[-]	[mm]	[-]	[-]
very stiff	-		C0	≥ 1.46
stiff	F1	≤ 340	C1	1.45 – 1.26
plastic	F2	350 – 410	C2	1.25 – 1.11
soft	F3	420 – 480	C3	1.10 – 1.04
very soft	F4	490 – 550	-	
flowable	F5	560 – 620	-	
very flowable	F6	≥ 630	-	

The consistency of modern quinary systems is also influenced by the additives and admixtures added to the concrete. The relationships between water/cement ratio, cement compressive strength and concrete compressive strength may vary significantly as a result of adding these constituents [1.30]. Additives and admixtures have a major effect on the workability characteristics of the concrete.

DIN EN 206-1:2001-07 [1.18] specifies several consistency ranges, which are summarised in Table 1.16. The tests most commonly carried out in Germany include slump tests and, for stiffer concretes, compacting factor tests as specified in DIN 1045-2:2008-08 [1.17]. The two resulting consistency classes cannot be directly related to each other.

The compacting factor is not suitable for very stiff, low-slump concretes. Such mixes have a low water/cement ratio and a very low cement paste content. They are thus considerably less compactable than standard concretes, which is why an increased amount of compaction energy must be used (preferably by imposing a load from above). This also applies to testing the consistency of the concrete mix, which is indispensable for efficient optimisation of the amounts of materials to be used.

Another option is the Proctor test. The Proctor density, determined in accordance with DIN 18127:1997-11 [1.37], is a soil mechanics parameter that is used to evaluate soil samples. This test involves impact compaction and can also be used to determine the optimal amount of water to be added to low-slump concrete mixes (Fig. 1.29). The diagram in Fig. 1.29 illustrates the influence of the amount of water added on the Proctor density whilst also clearly showing the effect of additives contained in the mix.

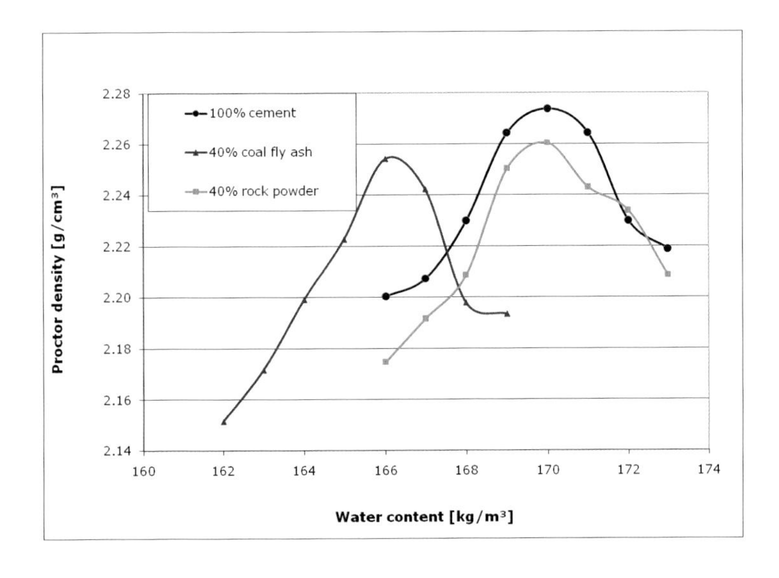

Fig. 1.29:
Dependence of Proctor density on the water content of the fresh concrete

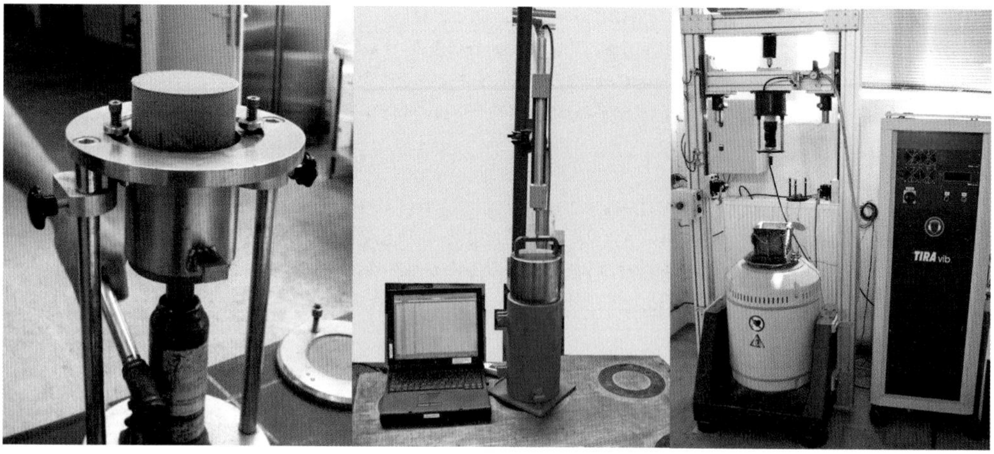

Fig. 1.30: Test equipment to determine the workability characteristics of very stiff concrete mixes
(left: Proctor device according to DIN 18127,
centre: "vibratory press compaction" set-up according to [1.41],
right: electrodynamic TIRAvib vibration test rig at IFF Weimar e. V.)

Low-slump concrete mixes are moulded by combined vibration and pressing – these two types of action ensure appropriate compaction of the mix. Hydraulically operated tamper systems generate the required loading pressure. By modification of the Proctor method and inclusion of vibratory compaction, the testing conditions were gradually adjusted to come as close as possible to those occurring during the manufacture of concrete products.

Comprehensive tests are described in [1.42]. The modified Proctor test developed in this publication is based on DIN 18127:1997-11 [1.37]. The bulk density of the fresh concrete is derived from the mass of the poured concrete mix and the volume measured after compaction.

The TIRAvib vibration test rig at IFF Weimar e. V. (Fig. 1.30, right) is a multi-purpose test set-up that permits evaluation of the workability characteristics of plastic to bulk material systems. Specimens to be tested can be subjected to excitation by selected waveforms, frequencies and accelerations. This rig implements all types of vertical and horizontal vibration (harmonic or pulsed vibration or a combination of the two) with or without a tamper head. The vibration test rig is used to evaluate the effect of excitation on a material system and to determine the vibration parameters (frequency, acceleration amplitude, compaction time) required to achieve the specified concrete properties. This system is also used to determine the bulk density of the fresh concrete as a function of the applied compaction energy.

b) Fresh concrete bulk density and air void ratio
When fresh concrete is compacted, the enclosed air bubbles escape, and the concrete aggregates achieve a higher packing density and are bound by the cement paste. The specified strength is reached, and the process results in a certain bulk density of the fresh concrete.

The bulk density of fresh concrete is defined as the quotient of the mass and the volume of the compacted fresh concrete. During concrete design, the theoretical bulk density of the fresh concrete can be calculated from the bulk densities of the raw materials for a given air void ratio.

The air void ratio indicates the degree of compaction. The void space in the fresh concrete is the residual space that remains after compaction. These air voids generally result from the compaction process because the loose concrete mix was not compacted fully during its placement and moulding. It is virtually impossible to eliminate all voids in the mix. Full compaction is difficult to achieve and usually requires prolonged and thus uneconomical vibration, which could cause unwanted segregation. Depending on the consistency of the mix, the ratio of air voids in the fresh concrete may be less than 2 vol.-%. This value increases with the stiffness of the mix (in general, the following values are found: class F2 – 2 vol.-%; C1 – 2.5 vol.-%; C0 – 3 vol.-%; i.e. between 20 and 30 l per m^3 of fresh concrete). The properties of such a concrete deteriorate only to a minor extent compared to a fully compacted concrete.

The air void ratio can also be modified by introducing artificial air voids. Concrete mixed with air-entraining agents or microspheres is called air-entrained concrete.

These constituents are added to create additional expansion space for water as it freezes. Artificially introduced air voids interrupt the largely continuous capillary pore system and reduce the amount of liquids absorbed by the concrete. The air voids are spherical and have diameters from 10 to 300 µm. They must be separated by a certain distance. Concretes containing these spheres have a greater resistance to frost and freeze-thaw cycles. On the other hand, the resulting voids adversely affect the compressive strength of the concrete, which is reduced by 1.5 to 2 N/mm^2 for each percent of added air [1.32].

c) Green strength
The concrete has a certain resistance to loading or deformation immediately after its placement and demoulding owing to the water film adhering to the solid constituents of the fresh concrete. This resistance is termed "green strength". At this stage, the cement hydration process has only just begun, if at all, which means that there is no chemical binding yet. The green strength depends on the strength parameters of the fresh concrete (green compressive strength) and on the shape and size of the items

produced. It is mainly determined by the water and cement contents, the particle size distribution, the shape of the aggregates and the amount of compaction energy applied.

The concrete has virtually no green compressive strength at high water ratios, which make fresh concrete plastic or soft. If the water ratio is reduced further, the concrete becomes stiffer and its green compressive strength increases. This strength reaches a maximum level at a particular water ratio, which depends on the degree of compaction. As the water ratio is reduced even further, the green strength then continues to decrease because the concrete becomes so stiff and hard to compact that it is no longer possible to create a coherent microstructure (Fig. 1.31).

A high cement ratio also has a favourable effect on the green compressive strength. As mentioned initially, however, this effect is not due to the increase in strength caused by the cement but to the change in the compactibility of the mix. In addition, finer cements lead to a minor increase in green compressive strength if all other conditions remain unchanged. The same applies to other ultrafine materials.

Another highly influential factor is the amount of compaction energy introduced: higher compaction energies allow processing of stiffer concrete mixes. As the degree of compaction increases, the maximum strength not only shifts to lower water ratios, it also increases at the same time.

The use of crushed, sharp-edged aggregates (chippings) instead of naturally rounded aggregates (gravel) leads to a more effective interlocking of the aggregates, thus increasing the green compressive strength. Aggregate mixes with low fines ratios and large maximum particle diameters also have a positive effect on the green compressive

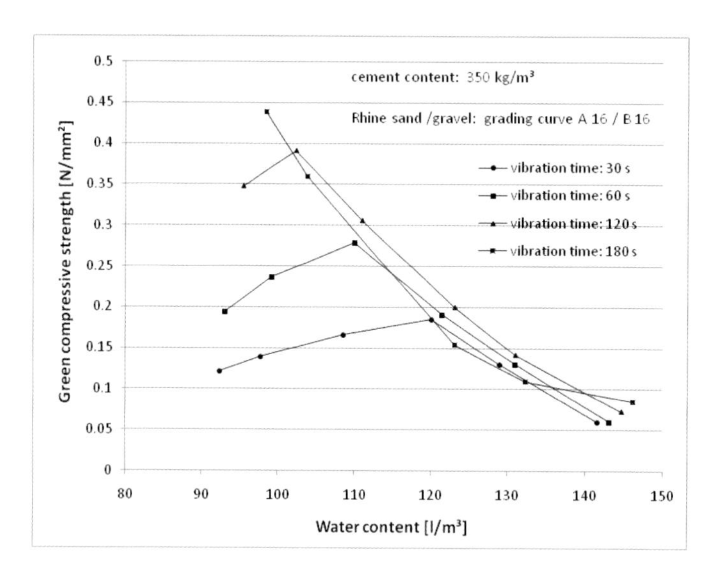

Fig. 1.31:
Green compressive strength as a function of the water content and compaction time (Wierig [1.31])

Fig. 1.32: Test specimen after measurement of its green compressive strength

strength. In contrast, increasing the fine sand ratio above a certain proportion requires the addition of larger amounts of water and reduces the green compressive strength. For this reason, favourable grading curves combine the highest possible packing density of the available aggregates with a continuous particle size distribution.

As described above, the green compressive strength depends on a large number of factors. Values between 0.1 N/mm^2 and 0.5 N/mm^2 can be achieved with stiff mixes.

1.2.3.2 Testing of the concrete mix/fresh concrete

There are a number of methods for testing the consistency that are more or less suitable for practical application. DIN 1045-2:2008-08 [1.17] specifies that the consistency of the concrete mix is to be determined either by the slump test in accordance with DIN EN 12350-5:2000-06 [1.35] or by the compaction test referred to in DIN EN 12350-4:2000-06 [1.34].

The slump test is suitable for the characterisation of concretes with a soft-to-flowable consistency. DIN EN 12350-5:2000-06 [1.35] specifies the procedure to be followed. Reference [1.33] also includes a description. This test simulates the reshaping of a portion of the concrete mix poured into a truncated cone to produce a concrete cake by means of a defined shock impact. The slump is the diameter of the concrete cake measured after 15 shocks (Fig. 1.33). The concrete must also be checked for segregation. The test must be repeated at defined intervals after mixing of the concrete in order to determine its working time.

Fig. 1.33: Slump test in accordance with DIN EN 12350-5:2000-06 [1.35]

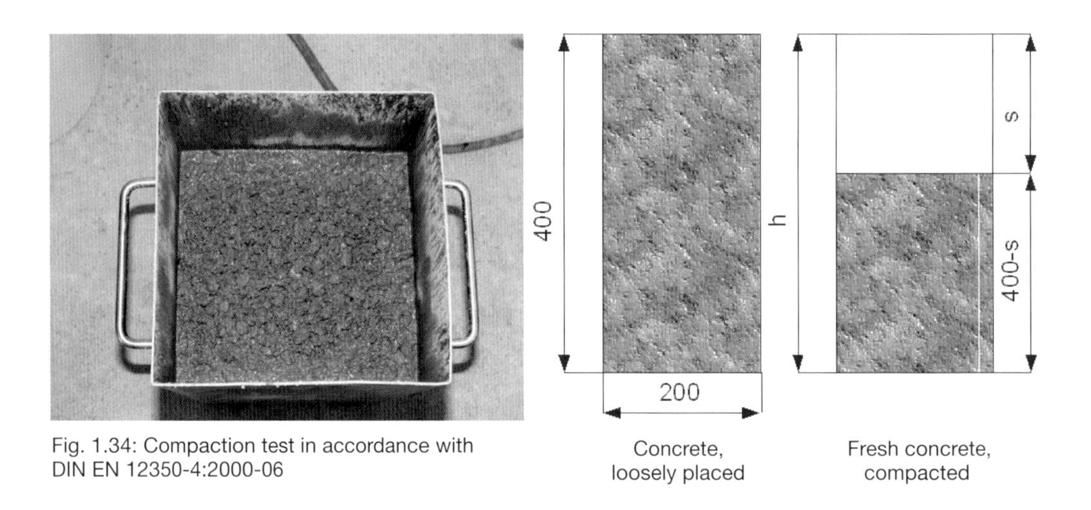

Fig. 1.34: Compaction test in accordance with DIN EN 12350-4:2000-06

The compaction test is used for plastic-to-stiff concrete mixes. DIN EN 12350-4:2000-06 [1.34], [1.33] specifies the procedure. In this test, a concrete mix is loosely poured into a container and then compacted. The change in height and volume is measured.

The air void ratio of compacted fresh concrete is determined with an air void test vessel using the pressure gauge method (Fig 1.35). DIN EN 12350-7:2000-11 [1.36] describes the procedure.

The concrete mix is placed in the vessel layer by layer and then compacted. Care must be taken to ensure that the concrete volume placed in the test vessel is exactly the same as the vessel volume. Removal of excess material should be avoided. After

Fig. 1.35: Determination of the air void content of fresh concrete

closing the upper part, water is pressed into the vessel. The pressure is then increased using the integrated air pump. The test key is then pressed and the pressure gauge will eventually display a stable value. Reference [1.33] also includes a description. The pressure gauge method is not suitable for bulk concrete mixes.

The bulk density of the fresh concrete is usually determined in the air void test vessel after completion of the compaction process. It is also possible to measure this parameter when producing specimens for compressive strength testing. The mass of the concrete contained in the mould can be determined with a maximum deviation of 0.1 % using the difference between the weights of the filled and empty mould. The volume of the test vessel is either known or must be determined by calibration. The bulk density is then calculated by dividing the concrete mass by the vessel volume, and is given with a maximum deviation of 10 kg/m^3.

1.2.3.3 Properties of hardened concrete

a) Concrete strength

The specifications for the hardened concrete are based on the mechanical loads and restraints as well as the chemical and physical loads applied to the structural concrete component.

The properties of the hardened concrete are essentially determined by the composition of the cement paste matrix. Key factors that influence this composition are the cement strength class, the water/cement ratio and the degree of hydration. The characteristics of the aggregate packing and the bond between the matrix and the aggregate are also of significance in this regard.

The strength of the concrete is a measure of the resistance of the concrete block to mechanical loads that cause deformation or separation. As the most important design parameter, it is determined in load-deformation tests carried out on concrete specimens. The following strength categories are used to characterise concrete:

- compressive strength
- axial tensile strength
- tensile bending strength
- tensile splitting strength
- adhesive tensile strength

The 28-day compressive strength is the key parameter for evaluating the concrete strength. Another important parameter in some applications is the early strength achieved after a few hours or days, e.g. the demoulding strength at the precast plant. The same applies to age hardening if a certain strength is required at a later concrete age. The 28-day compressive strength is also the basis for the definition of concrete strength classes contained in DIN EN 206-1:2001-07 [1.18] and DIN 1045-2:2008-08 [1.17].

The most significant factors that influence the development of the compressive strength are the paste matrix, the aggregate and the contact zone between the matrix and the aggregate.

The contribution of the aggregates to the concrete strength is determined by the selected grading curve as well as the particle strength, shape and surface. In the case of cement, these factors include the cement grade, strength class and quantity added.

Whereas the compressive strength of cement is determined by the clinker phase and milling fineness of the cements used in the mix, the water/cement ratio defines the void spaces that form within the paste. The w/c ratio influences the capillary volume and thus the strength and impermeability of the paste, which has an effect on the durability of the hardened concrete. A water/cement ratio of at least 0.4 is necessary to trigger complete hydration. Of this 40 M.-% water (relative to the cement content), 25 M.-% is bound chemically in the hydration products. The remaining 15 M.-% is bound physically in the gel pores. Any excess water still present after these reactions leads to capillary pores that reduce the concrete strength and transport liquids and gases. For this reason, DIN EN 206-1:2001-07 [1.18] and DIN 1045-2:2008-08 [1.17] introduce limits for the maximum water/cement ratios in the individual exposure classes. Concretes with low w/c ratios develop their strength more quickly, which is why they have a higher strength after 28 days.

The axial tensile strength refers to the mean tensile stress that an axially tensioned specimen is able to resist. DIN 1045-1:2008-08 [1.46] specifies that this parameter may be derived approximately from the splitting tensile strength (applying a factor of 0.9). The axial tensile strength amounts to only 5 to 10% of concrete compressive strength and cannot be used as a material parameter to describe the strength of the concrete.

Fluctuations in the concrete aggregates, types of loading and shape of the specimen have a more significant influence on the measured tensile strength than on the compressive strength.

Of all the relevant parameters, the axial tensile strength comes closest to the actual tensile strength of the concrete. However, its measurement requires a sophisticated set-up and is thus carried out very rarely. Instead, tensile bending and splitting strengths are determined in tests.

The adhesive tensile strength is used as a parameter to characterise the adhesion of layers, such as renders, screeds, coatings or paints, to the concrete surface. This type of strength is determined in an adhesive tensile test whose technical set-up and procedure do not differ from those used for tensile strength testing on concrete surfaces. In contrast to the adhesive tensile test, the surface tensile strength test determines cohesion in the concrete surface zone [1.43]. Reference [1.45] describes the testing method as part of concrete repairs. DIN EN 13813:2003-01 [1.44] specifies testing methods for screed mortars and screeds, whereas DIN EN 1015-12:2000-06 [1.38] governs tests for masonry mortars.

b) Deformation behaviour
In this category, a distinction must be made between load-induced deformation and deformation that occurs independently of any applied load.

Types of deformation that occur irrespective of any loading include deformation caused by temperature fluctuations as well as shrinkage and swelling. In the case of restrained components, in particular, these phenomena must not be neglected because they increase the risk of cracking.

Shrinkage processes lead to a decrease in the volume of the cement paste or concrete/mortar. The following types of shrinkage occur in the cement paste:

- plastic shrinkage (early shrinkage) is a decrease in volume that occurs prior to the onset of hardening. It is caused by drying triggered by exposure of the concrete to wind, sunlight, high temperatures and/or low humidity. This type of shrinkage is caused by dehydration as a result of capillary forces and is characterised by cracks that run perpendicular to the surface.
- contraction is the sum of chemical and autogenous shrinkage; the former is caused by hydration processes and the associated binding of water; the latter results from a decrease in volume due to internal withdrawal of free water as hydration progresses.
- drying shrinkage occurs due to the loss (evaporation) of excess water that is not bound chemically or physically; this process depends on the ambient conditions and occurs in hardened concrete.

- carbonation shrinkage is the reaction of atmospheric carbon dioxide with the calcium hydroxide in the cement paste; this process is irreversible and may result in a reticular pattern of surface cracks [1.28].

Permanent storage in a moist environment prevents significant shrinkage under practical conditions. Swelling processes are triggered if water is incorporated into the paste structure [1.28]. In addition, temperature-induced expansion may occur depending on the coefficients of thermal expansion of the aggregate and cement paste, the temperature difference and the moisture content of the concrete.

Changes in volume that occur due to plastic deformation under load are referred to as "creep". Creep depends on the magnitude and period of loading, the ambient conditions, the water/cement ratio, the cement content, the type of aggregate used and the degree of concrete hardening at the time of loading.

Load-induced elastic deformation is reversed when the load is no longer applied. The main factors that influence elastic deformation are the type of aggregate, concrete strength, water/cement ratio, storage conditions and age. Elastic deformation is the quotient of strain and the modulus of elasticity of the concrete. The modulus of elasticity is a material parameter that describes the correlation between strain and expansion during the deformation of a solid, assuming linearly elastic behaviour. The greater the resistance to deformation that the material exhibits, the higher its modulus of elasticity, and the higher the modulus of elasticity, the lower the degree of deformation of the test specimen under load.

Any longitudinal deformation is associated with a transverse deformation in the opposite direction. For instance, compressive stresses acting on the specimen result in transverse strain. The ratio of transverse to longitudinal strain is termed "coefficient of transverse strain" (or Poisson's ratio).

c) Durability

Durability refers to retention of the performance characteristics over the intended service life under the loads and stresses provided for in the design whilst also considering cost efficiency (low maintenance costs).

From a material science point of view, this concept refers to the resistance of the building material to environmental impact. Apart from strength, [1.32] specifies a number of direct durability indicators, which are summarised in Table 1.17.

A high durability can be achieved if all relevant rules and standards are adhered to, and if the following conditions are fulfilled:

- verification of the specified concrete compressive strength
- selection of aggregates with an appropriately adjusted gradation and high packing density

Table 1.17: Durability indicators [1.32]

Direct indicators	Verification
Impermeability	water penetration depth under pressure water absorption in contact with water on one or all sides gas permeability
Frost resistance/ freeze-thaw resistance	loss of mass loss of volume expansion behaviour change in the modulus of elasticity
Carbonation depth/ chloride penetration depth	indicators test reaction
Effect of aggressive fluids	strain measurements loss of volume loss of strength change in the modulus of elasticity light and electron microscopy
Alkali-silica reaction	strain and crack formation (cloud chamber test) light and electron microscopy

- use of concrete additives and admixtures, such as concrete workability agents (CWA), plasticisers (P), air-entraining agents (AEA), fly ash
- implementation of an optimised mixing and compaction process
- sufficient post-treatment [1.32]

DIN EN 206-1:2001-07 [1.18] and DIN 1045-2:2008-08 [1.17] specify the following criteria for durable external structural components:

- concrete compressive strength
- maximum permissible water/cement ratios
- minimum cement quantities
- minimum concrete covers
- maximum permissible void space
- maximum permissible crack widths
- use of air-entraining agents with a minimum air content in the fresh concrete

These criteria must be defined in relation to the relevant exposure class.

To produce a durable concrete, it is crucial to achieve a dense packing by means of a low water/cement ratio and to ensure optimal compaction and a sufficiently long post-treatment period.

The durability of the concrete for the intended use under the prevailing local conditions is deemed verified if the concrete complies with these specifications. It is assumed that

- the appropriate exposure and moisture classes have been selected
- the concrete has been poured and compacted in accordance with applicable rules and standards

- the minimum concrete cover has been adhered to
- appropriate maintenance measures have been taken

1.2.3.4 Testing of hardened concrete

Testing of hardened concrete requires the preparation of appropriate specimens, which is regulated by DIN EN 12390-2:2009-08 [1.82]. The National Annex to this standard describes the storage conditions for compressive strength and tests of the modulus of elasticity in Germany. Dry storage must be used in accordance with the following steps:

- the prepared specimens must be stored for 24 ± 2 hours at 20 ± 2 °C and protected against drying
- demoulding after 24 ± 2 hours
- demoulded specimens must be stored for 6 days at 20 ± 2 °C on grates in a water bath or on a grid in a humidity chamber at > 95% relative humidity
- from the age of 7 days to the test date, the specimens must be stored at 20 ± 2 °C and 65 ± 5% relative humidity.

Cylinders, cubes or core samples are used for compressive strength testing in accordance with DIN EN 12390-3: 2009-07 [1.83]. Compressive strength is the quotient of the maximum load at failure and the area of the sample cross-section. It is stated in N/mm^2. The compressive strength after storage in water (reference storage) is used to assign a strength class to the concrete as specified in DIN EN 206-1:2001-07 [1.18]/ DIN 1045:2008-08 [1.17]. Whereas dry storage was applied to the test specimens in accordance with the National Annex to DIN EN 12390-2:2009-08 [1.82], the strength determined in the test must be stated with respect to the reference storage parameters.

Bar-shaped specimens with a square cross-section are used for tensile bending tests, which are regulated by DIN EN 12390-5:2009-07 [1.85]. These tests involve either a two-point loading or axial loading arrangement. The test specimen is bent to failure.

Fig. 1.36: Examples of compressive strength and tensile bending strength testing

The quotient of the ultimate moment and the section modulus is calculated. The tensile bending strength is stated in N/mm². The selected testing method must always be stated because the tensile bending strengths achieved are usually 13% higher if axial loads are applied.

Concrete cylinders are used to determine the splitting tensile strength in accordance with DIN EN 12390-6:2001-02 [1.86]. The concrete cylinder is loaded to failure between two linear contact points on opposing surfaces. The splitting tensile strength is derived from the ultimate load and the dimensions of the test specimen and is stated in N/mm².

According to DIN EN 13813:2003-01 [1.44], the adhesive tensile strength is determined by pulling off a test disc that has been adhesively bonded to the coating of the test specimen under defined conditions (measuring area, temperature, pull-off velocity etc.) using a pull-off testing rig. The test disc is pulled perpendicular to the concrete surface at a slow and constant rate until it breaks off (failure). In addition to the measured value, the description of where failure occurred (where the item broke off) is another key criterion. The measured adhesive tensile pull-off strength can never be greater than the inherent strengths of the individual components. In a bond consisting of several components, the weakest link is always the determining factor (Fig. 1.37).

Several standards specify the testing methods used to determine the static and dynamic moduli of elasticity, and thus to characterise the deformation behaviour of concrete.

The static modulus of elasticity is measured in a compression tester in accordance with DIN 1048-5:1991-06 [1.40]. For this purpose, a test cylinder with flat and parallel end surfaces is marked with measuring distances on symmetrical surface lines. Changes in these distances are then measured under load and again after unloading. Hardened concrete prisms and core samples are mainly used for this type of test. Test specimens with an approximate length ratio of 1:3 are required because the initial points of

Fig. 1.37: Specimen after the adhesive tensile test and a pull-off tester

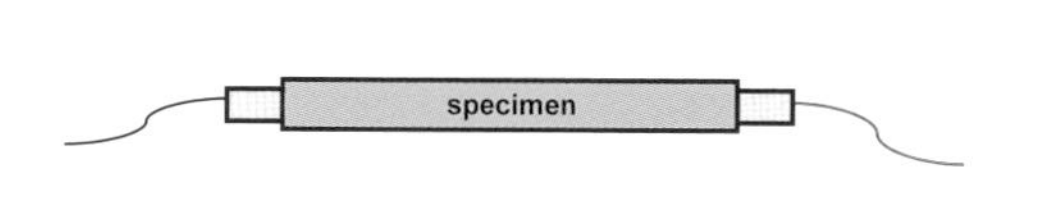

Fig. 1.38:
Test rig to measure the resonance frequency

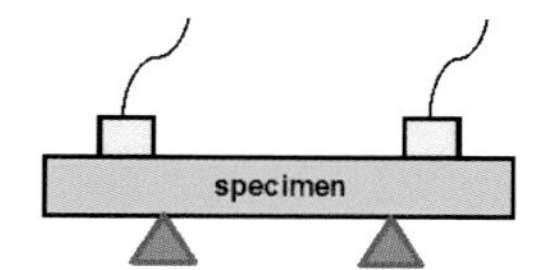

Fig. 1.39:
Test rig to measure flexural vibration

the measuring distances must be located at a certain minimum distance from the end surfaces.

The dynamic modulus of elasticity is determined in a non-destructive test. It is calculated from the bulk density and the velocity at which a mechanical momentum propagates through a specimen (ultrasonic velocity) in accordance with DIN EN 12504-4:2004-12 [1.47]. This method is particularly suitable for determining this modulus quickly and easily with specimens whose properties are expected to change over time and thus need to be identified.

The first resonance frequency is measured by longitudinal excitation of a bar-shaped specimen and is also determined for transverse excitation. Longitudinal excitation (Fig. 1.38) yields the longitudinal wave velocity as an intermediate result – a material parameter that is required to calculate the modulus of elasticity. In the transverse excitation mode (Fig. 1.39), the bar is excited laterally to perform flexural vibration. The dynamic modulus of elasticity is then calculated from the first flexural resonance frequency. Transverse vibration can be applied if the material is fine-grained, dense and solid. The velocity at which this vibration propagates through the specimen is used to calculate the dynamic shear modulus (G).

Table 1.17 lists the parameters that are used to verify durability, which can be determined using the test methods described below.

The verification of the water penetration depth, specified in DIN EN 12390-8:2009-07 [1.39], is a method to test the impermeability of the hardened concrete. This involves applying water to the surface of the test specimen at a certain pressure (500 kPa). After this compressive action, the specimen is split in order to measure the greatest penetration depth, which is stated with an accuracy of 1 mm. Specimens must be at least 28 days old when the test commences (Fig. 1.40).

The frost and freeze-thaw resistance of hardened concrete is determined in accordance with DIN CEN/TS 12390-9:2006-08 [1.49]. This standard contains a reference test method and two alternative procedures used to determine the degree of surface scaling.

Fig. 1.40:
Test of water penetration depth

Reference test method (slab test): measurement of the frost resistance involves exposing prism-shaped specimens (slab dimensions: 150 mm x 150 mm x 50 mm) to demineralised water. To determine their freeze-thaw resistance, the slabs are exposed to a 3% sodium chloride solution. Each freeze-thaw cycle lasts 24 hours. The mass of the scaled material is determined and stated in kg/m^2.

Alternative testing method (cube test): cube-shaped specimens (side length: 100 mm) are fully immersed in demineralised water to determine their frost resistance and in a 3% sodium chloride solution to determine their freeze-thaw resistance. The specimens are exposed to 7, 14, 28, 42, and 56 freeze-thaw cycles; each of these cycles lasts 24 hours. The material that has scaled off the entire surface of the specimen is collected, its mass measured and the loss of mass calculated as a percentage. The determining parameter is the loss of mass after 56 freeze-thaw cycles.

Alternative testing method (CF/CDF test): prism-shaped specimens (slab dimensions: 150 mm x 150 mm x 70 mm) are immersed in demineralised water (CF test) or in a 3% sodium chloride solution (CDF test) in such a way that the test surface is located at the bottom. The specimens are then subjected to 14, 28, 42 and 56 freeze-thaw cycles in the CF test and to 4, 6, 14 and 28 cycles in the CDF test. Each freeze-thaw cycle lasts 12 hours. The mass of the material that has scaled off the test surface is determined and stated in kg/m^2. The determining parameter is the total value after 56 (CF test) or 28 (CDF test) freeze-thaw cycles.

The carbonation depth is determined on the basis of [1.48] for fresh fracture surfaces of the concrete to be tested. For this purpose, an indicator solution consisting of phenolphthalein is sprayed onto these surfaces. Non-carbonated areas appear in red whereas the carbonated area remains unchanged. The carbonation depth refers to the

Fig. 1.41:
Measurement of the carbonation depth in concrete samples

maximum distance (stated in mm) between the coloured zone and the external surface of the concrete (Fig. 1.41).

Concrete is damaged by the alkali-silica reaction if the volume increases triggered by the reaction lead to stresses that exceed the tensile strength of the concrete. In such a case, cracking and spalling occur. [1.32] deals with this topic in detail. Various test methods are used around the world to assess the sensitivity of mineral aggregates to alkali-silica reactions. In Germany, the Alkali-Richtlinie (Alkali Guideline) [1.50] describes the procedure to be followed for the assessment of aggregates.

This guideline states that the aggregate should first be subjected to an initial petrographic evaluation and then to initial testing, if required. On the basis of the test result, the aggregate is allocated to one of the alkali sensitivity classes defined in the guideline. Test aggregates with particle sizes from 1 to 4 mm are exposed to a hot 4% sodium hydroxide solution, and their mass loss is determined. Sizes greater than 4 mm are separated into fractions that are clearly insensitive to alkaline attack and into flint, opaline sandstone, siliceous chalk and unidentified constituents. The opaline sandstone, siliceous chalk and the unidentified constituents are then exposed to a hot 10% sodium hydroxide solution, and their mass loss is measured. Crushed, alkali-sensitive aggregates can be subjected to a quick test (reference test) and/or the concrete test in a fog chamber.

The quick test is performed on mortar prisms (40 mm x 40 mm x 160 mm) that contain the aggregates to be tested. The specimens are stored in a sodium hydroxide solution heated to 80 °C for a defined period. The expansion of the prisms is measured.

In the fog chamber test, which is carried out at a temperature of 40 °C for a period of nine months, the expansion pattern of concrete prisms (100 mm x 100 mm x 500 mm) is investigated. The change in length is recorded at regular intervals. A cube (side length: 300 mm) is also stored and used to observe any cracking.

1.3 Product Fundamentals

1.3.1 Concrete Products

This chapter deals with the following three product groups:

- small concrete products
- precast concrete elements
- concrete pipes and manholes

All these products are prefabricated at the factory.
Small concrete products can again be sub-divided into several categories, one of which comprises concrete products for road construction. These mainly include mass-produced small items without a significant structural function (Fig. 1.42). Most of these products are demoulded immediately after casting. They are divided into:

- concrete paving blocks in accordance with DIN EN 1338:2003-08 [1.65]
- concrete paving flags in accordance with DIN EN 1339:2003-08 [1.66]
- concrete kerbstones in accordance with DIN EN 1340:2003-08 [1.67]

Concrete masonry units and roof tiles are also regarded as small concrete products. The specifications for masonry units (Fig. 1.43) are given in DIN EN 771-3:2005-05 [1.69] together with the applicable preliminary standards DIN V 18151-100:2005-10 [1.70]

Fig. 1.42: Left: paving blocks; centre: paving flags; right: concrete kerbs

Fig. 1.43: Concrete masonry units

Fig. 1.44: Examples of various types of concrete masonry units

Fig. 1.45: Cast stones

(lightweight concrete hollow blocks), DIN V 18152-100:2005-10 [1.72] (lightweight concrete solid bricks and blocks) and DIN V 18153-100:2005-10 [1.72] (concrete masonry units, normal-weight concrete).

Fig. 1.44 shows several types of masonry units.

The product specifications for concrete roof tiles and fittings for roof covering and wall cladding are given in DIN EN 490:2006-09 [1.73]. The associated test methods are defined in DIN EN 491:2005-03 [1.74].

Cast stones are made of reinforced or non-reinforced concrete containing cement and mineral aggregates. After prefabrication, their surfaces are finished by applying stone-masonry techniques or using special, textured formwork. There are very diverse options to design the surface of cast stones, which is why they can be used for a wide range of applications. The main product groups are terrazzo tiles, steps and step coverings, façade panels and other items. DIN V 18500:2006-12 [1.75] includes the related product specifications and test methods.

Precast concrete products in accordance with DIN EN 13369:2004-09 [1.76] are structural components made of concrete or reinforced or prestressed concrete that are designed on the basis of the appropriate product standard or DIN EN 13369:2004-09 [1.76] and manufactured in a location other than their place of final assembly.

Fig. 1.46: Precast concrete elements

Fig. 1.47: Concrete pipes

Precast concrete products cover a comprehensive and extremely diverse range of prefabricated concrete items, most of which have large dimensions and are used as load-bearing elements in building construction and civil engineering. Examples include wall and floor units, columns, beams, girders and box units (Fig. 1.46).

Concrete pipes in accordance with DIN EN 1916:2003-04 [1.77] and DIN V 1201:2004-08 [1.78] are hollow prefabricated elements made of concrete, reinforced or steel-fibre reinforced concrete. They are used to transport wastewater, stormwater and surface water. They are produced with or without base and with a uniform internal cross-section across their entire length (with the exception of the connecting sections). The connecting parts of these components are pre-formed as spigots and sockets and include one or several seals (Fig. 1.47).

Concrete manholes in accordance with DIN V 4034-1:2004-08 [1.80] and DIN EN 1917:2003-04 [1.79] are structures designed to connect to buried sewers or sewage pipes. They are mainly used for ventilation, inspection, maintenance and cleaning purposes and may also include systems for elevating the level of the wastewater, enabling the merger of pipelines, or changing the direction, gradient or cross-section of sewers and pipelines. Manholes usually consist of precast concrete items with socket fittings (Fig. 1.48, Fig. 1.49).

In addition, there are a number of non-standardised concrete products, such as grass pavers and concrete products for bulk containers. The specifications of their product

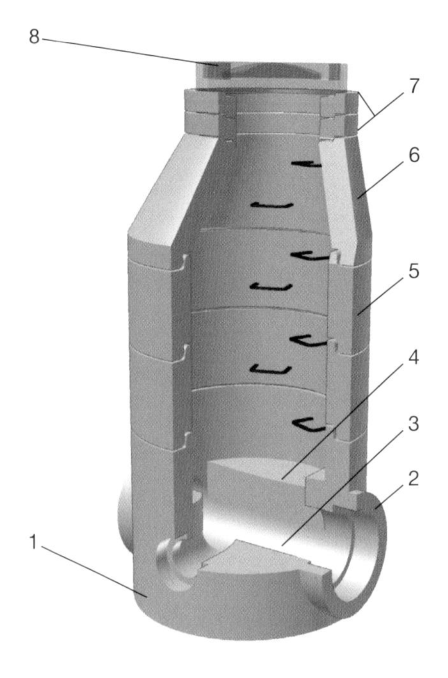

Fig. 1.48: Example of a manhole made of precast concrete and reinforced concrete components in accordance with DIN V 4034-1:2004-08 [1.80]
1 Manhole base
2 Connecting element
3 Channel
4 Tread
5 Manhole ring
6 Manhole neck (cone)
7 Top ring
8 Manhole cover according to DIN EN 124

properties were defined by the Bund Güteschutz Beton- und Stahlbetonfertigteile e. V. (BGB; Association for the Quality Assurance of Precast Concrete and Reinforced Concrete Elements) as part of the BGB guideline pertaining to "Nicht genormte Betonprodukte – Anforderungen und Prüfungen – (BGB-RiNGB)" (Non-standardised Concrete Products – Specifications and Tests). This guideline was last updated in 2005.

Fig. 1.49: Precast concrete manhole elements

1.3.2 Requirements Relating to Product Characteristics and Testing Methods

1.3.2.1 Requirements for small concrete products

a) Concrete products for road construction

At the European standardisation level, the requirements relating to the product characteristics are divided into several classes (qualities). This system serves as the basis for

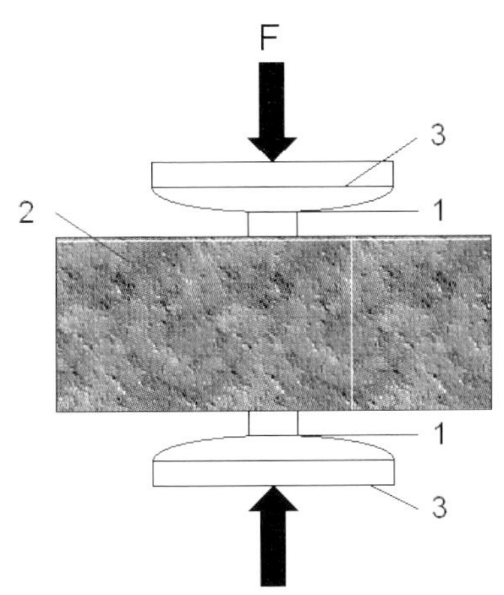

Fig. 1.50: Principle of splitting tensile strength testing in accordance with DIN EN 1338:2003-08 [1.65]
1 Load distribution strip
2 Paving block
3 Rigid load blades

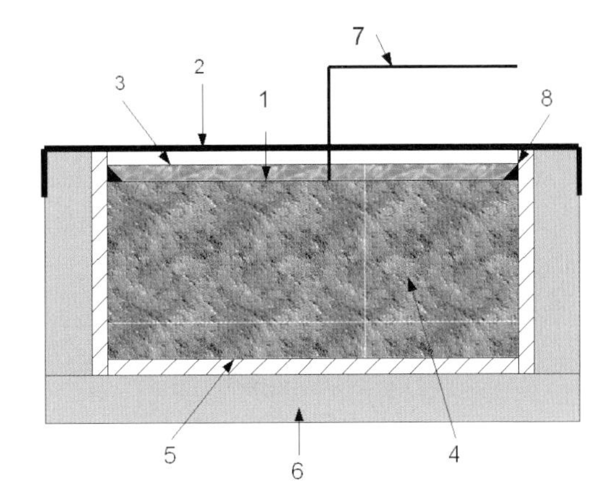

Fig. 1.51: Basic set-up for a freeze-thaw test in accordance with DIN EN 1338:2003-08 [1.65]
1 Test surface
2 Polyethylene film
3 De-icing salt solution
4 Test specimen
5 Rubber layer
6 Thermal insulation
7 Temperature gauge
8 Sealant

the individual EU member states to select a certain class for a defined product requirement from the standard in order to determine the requirements specific to the country and implement them in a set of rules for national application [1.51].

In Germany, the product characteristics that govern the use of concrete paving blocks, paving flags and kerbstones were defined in the new "Technische Lieferbedingungen für Bauprodukte zur Herstellung von Pflasterdecken, Plattenbelägen und Einfassungen" (Technical Specifications for Construction Products to Lay Block Pavements, Slab Pavements and Kerbs), 2006 edition, FGSV-Verlag (TL-Pflaster-StB 06) [1.52].

The general specifications for concrete products relate to their:

- quality
- shapes and dimensions
- mechanical strength
- abrasion resistance
- sliding and slip resistance
- weather resistance

The most important properties to characterise paving blocks are their tensile splitting strength, weather resistance and wear resistance.

The previously tested parameter of compressive strength was replaced by tensile splitting strength as a result of the introduction of the European standard. Fig. 1.50

Fig. 1.52: Freeze-thaw resistance test (paving block prior to and after frost test)

illustrates the test principle. The paving block must be placed in the tester in such a way that the load distribution strips, which are in contact with the load blades, are located at the top and the bottom of the block. Splitting patterns must be selected in accordance with DIN EN 1338:2003-08 [1.65]. The load must be applied uniformly and in increments of (0.05 ± 0.01) MPa/s; the failure load must be documented.

Weathering resistance is determined by a freeze-thaw cycle test using de-icing salt. For this purpose, a pre-conditioned test specimen, whose surface was previously treated with a 3% sodium chloride solution, is exposed to 28 freeze-thaw cycles. The spalled material is collected and weighed, and the results stated in kg/m². This test method is also used for paving flags in accordance with DIN EN 1339:2003-08 [1.66] and kerbstones specified in DIN EN 1340:2003-08 [1.67].

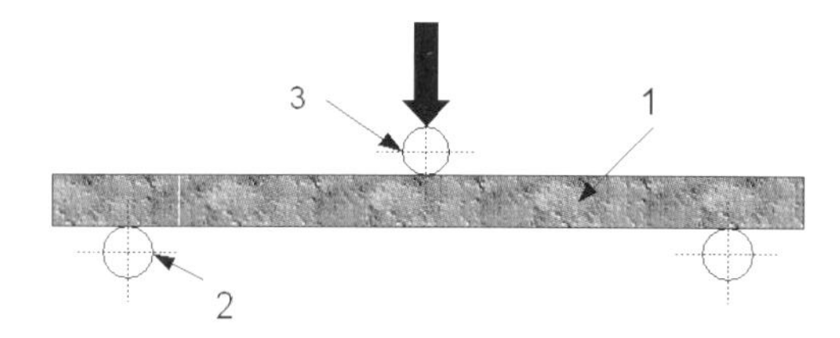

Fig. 1.53:
Principle of the tensile bending strength test of paving flags in accordance with DIN EN 1339:2003-08 [1.66]
1 Test flag
2 Support
3 Load blade

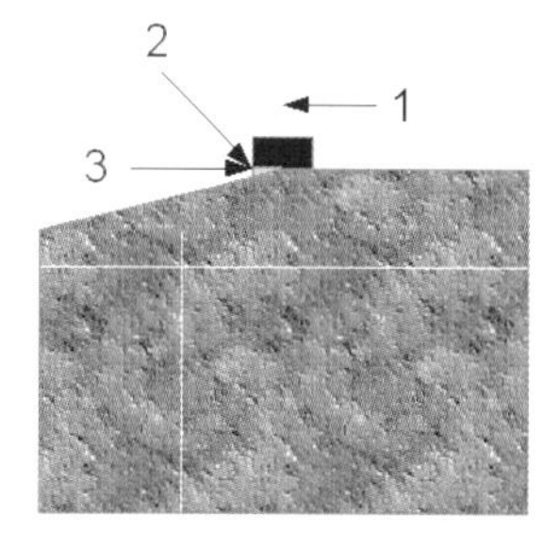

Fig. 1.54:
Principle of the tensile bending strength test of kerbstones in accordance with DIN EN 1340:2003-08 [1.67]
1 Neutral axis
2 Base
3 Hardwood wedge

For paving flags and kerbstones, tensile bending strength is tested as the primary parameter. In general, entire flags or kerbs are used. Paving flags may also be cut to size, but they must have two parallel, straight edges. Specimens must be stored in water for 24 ± 3 hours at a temperature of 20 ± 5 °C. They are subsequently taken out, dried and tested immediately thereafter. Once the paving blocks or kerbstones have been placed in the tester, as shown in Fig. 1.53 and 1.54, the load must be applied in defined increments. The failure load must be documented, and the tensile bending strength calculated.

Table 1.18: Specifications for paving blocks conforming to DIN EN 1338:2003-08 and TL-Pflaster-StB 06 [1.52], [1.51], [1.65]

Product characteristic	Requirement	Test methods
Dimensions	The length/thickness ratio must be ≤ 4. No size limit	DIN EN 1338 Annex C Measurement
Permissible deviations from nominal dimensions	Block thickness < 100 mm: length, width ± 2 mm; thickness ± 3 mm Block thickness > 100 mm: length, width ± 3 mm; thickness ± 4 mm	DIN EN 1338 Annex C Measurement
Evenness of surface[1)]	Convex deviation: ≤ 1.5 or ≤ 2.0 mm (depending on measured length) Concave deviation: ≤ 1.0 or ≤ 1.5 mm (depending on measured length)	DIN EN 1338 Annex C Measurement
Max. difference of both diagonals[1)] (squareness)	Class 2, Label K ≤ 3 mm	DIN EN 1338 Annex C Measurement
Mechanical strength	Tensile splitting strength ≥ 3.6 N/mm² (0.05 quantile) ≥ 2.9 N/mm² (single value) Each length-related failure load ≥ 250 N/mm	DIN EN 1338 Annex F Tensile splitting strength test
Abrasion resistance	Class 4, Label I ≤ 20 mm (reference method) or ≤ 18 cm³/50 cm² (Böhme test)	DIN EN 1338 The test may also be carried out in accordance with Annex H using the Böhme grinding wheel. The reference method, however, is the abrasion test with a wide grinding wheel, as specified in Annex G.
Sliding/slip resistance	Blocks have a sufficient sliding/slip resistance if they are not ground, polished or finished in such a way that a smooth surface has been created. The manufacturer must specify a minimum value for all other blocks.	DIN EN 1338, Annex I A pendulum device with specified characteristics must be used for the sliding/slip resistance test.
Weather resistance	Class 3, Label D Loss of mass after freeze-thaw test: ≤ 1.0 kg/m² (mean value) ≤ 1.5 kg/m² (single value)	DIN EN 1338, Annex D "Slab test" – the test specimen is pre-conditioned, its surface covered with a 3 % sodium chloride solution and exposed to 28 freeze-thaw cycles. The spalled material is collected and weighed, and the results stated in kg/m².

[1)] Applies exclusively to blocks above a certain size

Table 1.19: Specifications for paving flags conforming to DIN EN 1339:2003-08 and TL-Pflaster-StB 06 [1.54], [1.52], [1.66]

Product characteristic	Requirement	Test methods
Dimensions	The length/thickness ratio must be > 4. Maximum length 1 m.	DIN EN 1339 Annex C Measurement
Permissible deviations from nominal dimensions	Class 2, Label P Nominal dimension ≤ 600 mm: length ± 2 mm, width ± 2 mm, thickness ± 3 mm Nominal dimension > 600 mm: length ± 3 mm, width ± 3 mm, thickness ± 3 mm The difference between any two length, width and thickness measurements performed for a single paving flag must not exceed 3 mm.	DIN EN 1339 Annex C Measurement
	Convex deviation: ≤ 1.5 to ≤ 4.0 mm (depending on measured length) Concave deviation: ≤ 1.0 or ≤ 2.5 mm (depending on measured length)	DIN EN 1339 Annex C Measurement
Max. difference of both diagonals[1] (squareness)	Class 2, Label K ≤ 3 mm for diagonals ≤ 850 mm ≤ 6 mm for diagonals > 850 mm	DIN EN 1339 Annex C Measurement
Tensile bending strength	Class 3, Label U ≥ 5.0 N/mm^2 (0.05 quantile) ≥ 4.0 N/mm^2 (single value)	DIN EN 1339 Annex F Tensile bending test
Failure load	Class 30, Label 3 ≥ 3.0 kN (0.05 quantile); ≥ 2.4 kN (SV)[2] Class 45, Label 4 ≥ 4.5 kN (0.05 quantile); ≥ 3.6 kN (SV) Class 70, Label 7 ≥ 7.0 kN (0.05 quantile); ≥ 5.6 kN (SV) Class 110, Label 11 ≥ 11.0 kN (0.05 quantile); ≥ 8.8 kN (SV) Class 140, Label 14 ≥ 14.0 kN (0.05 quantile); ≥ 11.2 kN (SV) Class 250, Label 25 ≥ 25.0 kN (0.05 quantile); ≥ 20.0 kN (SV) Class 300, Label 30 ≥ 30.0 kN (0.05 quantile); ≥ 24.0 kN (SV)	DIN EN 1339 Annex F Tensile bending test
Abrasion resistance	Class 4, Label I identical to paving blocks	DIN EN 1339 Annex G or Annex H identical to paving blocks
Sliding/slip resistance	identical to paving blocks	DIN EN 1339, Annex I identical to paving blocks
Weather resistance	Class 3, Label D identical to paving blocks	DIN EN 1339, Annex D identical to paving blocks

[1] Applies exclusively to paving flags above a certain size
[2] SV – single value

For paving blocks conforming to DIN EN 1338:2003-08 [1.65], Table 1.18 lists the specifications and the test methods to be applied in order to determine their properties.

Table 1.20: Specifications for kerbstones conforming to DIN EN 1340:2003-08 and TL-Pflaster-StB 06 [1.55], [1.52], [1.67]

Product characteristic	Requirement	Test methods
Dimensions	Shapes and sizes may be defined nationally (DIN 483, April 2004 edition, specifies them).	DIN EN 1340, Annex C Measurement
Permissible deviations from nominal dimensions	Length ± 1%, rounded to full millimetres, min. ± 4 mm, max. ± 10 mm Dimensions of visible surfaces ± 3%, rounded to full millimetres, min. ± 3 mm, max. ± 5 mm Dimensions of other surfaces ± 5%, rounded to full millimetres, min. ± 3 mm, max. ± 10 mm The difference between any two measurements of a single dimension must not exceed 5 mm.	DIN EN 1340, Annex C Measurement
Evenness of surfaces and straightness of edges	Permissible deviation: ± 1.5 to ≤ 4.0 mm (depending on measured length)	DIN EN 1340, Annex C Measurement
Tensile bending strength	Class 2, Label T ≥ 5.0 N/mm² (0.05 quantile) ≥ 4.0 N/mm² (single value)	DIN EN 1340, Annex F Tensile bending test
Abrasion resistance	Class 4, Label I identical to paving blocks	DIN EN 1340, Annex G or Annex H identical to paving blocks
Sliding/slip resistance	identical to paving blocks	DIN EN 1340, Annex I identical to paving blocks
Weather resistance	Class 3, Label D identical to paving blocks	DIN EN 1340, Annex D identical to paving blocks

Table 1.19 lists the specifications for paving flags in accordance with DIN EN 1339:2003-08 [1.66] and the test methods that serve to determine their properties.

Table 1.20 contains the corresponding specifications and test methods for kerbstones conforming to DIN EN 1340:2003-08 [1.67].

b) Concrete masonry units

Masonry units are prefabricated elements used for the construction of both load-bearing and non-load-bearing external and internal masonry walls. Concrete masonry units are specified in DIN EN 771-3:2005-05 [1.97].

General specifications for concrete masonry units relate to their:

- quality
- shapes and dimensions
- block and concrete bulk density
- mechanical strength
- thermal insulation characteristics
- durability

Table 1.21: Specifications for concrete masonry units in accordance with DIN EN 771-3:2005-05 [1.97]

Product characteristic	Requirement	Test methods
Dimensions and dimensional limits	Mean values from six individual values Plane parallelism and total of block thicknesses from three individual values Declared values in mm and dimensional classes	DIN EN 772-16 [1.103] DIN EN 772-2 [1.104] Measurement
Shape and design	Declaration as in DIN EN 1996-1, either as a range of values or as upper and lower limits	DIN EN 772-16 [1.103] DIN EN 772-2 [1.104] DIN EN 772-20 [1.105] Measurement
Compressive strength	Specifications in DIN V 18151-100, DIN V 18152-100, DIN V 18153-100, Annex A Instead of compressive strength, the mean tensile bending strength of blocks with a width smaller than 100 mm and a length/width ratio greater than 10 may be specified by the manufacturer. Values must not be lower than the value declared in N/mm².	DIN EN 772-1 [1.106] DIN EN 772-6 [1.107]
Dimensional stability	Must be specified if required; declared value in mm/m must not be exceeded	DIN EN 772-14 [1.108] Humidity-induced expansion
Bond strength	Must be specified if required; values must not be lower than the value declared in N/mm².	DIN EN 1052-3 [1.112] Initial shear strength Adhesive tensile bending strength
Fire behaviour	Test not required if masonry units contain ≤ 1.0 per cent by mass or volume of evenly distributed organic constituents (Class A1).	DIN EN 13501-1 [1.113]
Durability	Must be specified if required.	Frost resistance (to be evaluated in accordance with rules and standards applicable in the place of use)
Water absorption	Must be specified if required; declared value in g/m²s must not be exceeded.	DIN EN 772-11 [1.110]
Water vapour transmission	Must be specified if required.	DIN EN ISO 12572 [1.114]
Airborne sound insulation	The manufacturer must state the gross dry bulk density of the units in kg/m³. The mean value of the tested specimens must not deviate by more than 10% from the declared value. Declared shape as described above.	DIN EN 772-13 [1.111]
Thermal insulation characteristics	Must be specified if required.	DIN EN 1745 [1.115]

Although the harmonised part of the DIN EN 771-3:2005-05 [1.97] European standard contains the characteristics of masonry units to be indicated in conjunction with CE marking, it does not cover all the specifications that apply to the use of masonry units in Germany in accordance with DIN 1053 [1.98]. For this reason, the preliminary standards DIN V 18151-100:2005-10 [1.99], DIN V 18152-100:2005-10 [1.100] and DIN V 18153-100: 2005-10 [1.101] specify these additional requirements.

c) Concrete roof tiles
DIN EN 490:2006-09 [1.73] contains the product specifications for concrete roof tiles and fittings whereas DIN EN 491:2005-03 [1.74] specifies the test methods.
General specifications for concrete roof tiles cover the following categories:

- quality
- shapes and dimensions
- structural strength
- water impermeability
- freeze-thaw resistance

Concrete roof tiles are primarily tested for structural strength and water impermeability in order to assess their characteristics.

To determine their structural strength, a load is applied to the roof tiles using a bending tester. The tile is placed face-up on the bending supports of the tester in such a way that its centre is located underneath the bending blade. For level, flat roof tiles, an elastic intermediate layer (elastomer base) must be inserted between the bending blade and the tile (see Fig. 1.55, top). For profiled roof tiles, a corresponding adjustment piece must be placed between the bending blade and the tile (Fig. 1.55, bottom).

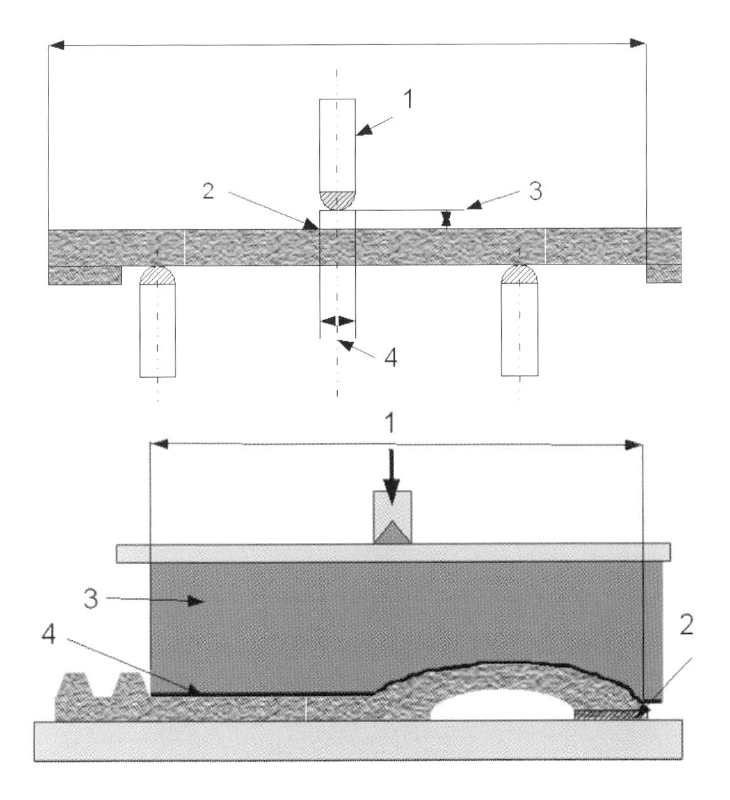

Fig. 1.55:
Strength test method
in accordance with
DIN EN 491:2005-03 [1.74]
1 Load
2 Elastomer base
3 10 mm ± 5 mm
4 ≥ 20 mm

1 Load
2 Adjustment pieces
3 Profiled hardwood or metal adjustment piece
4 Elastomer base

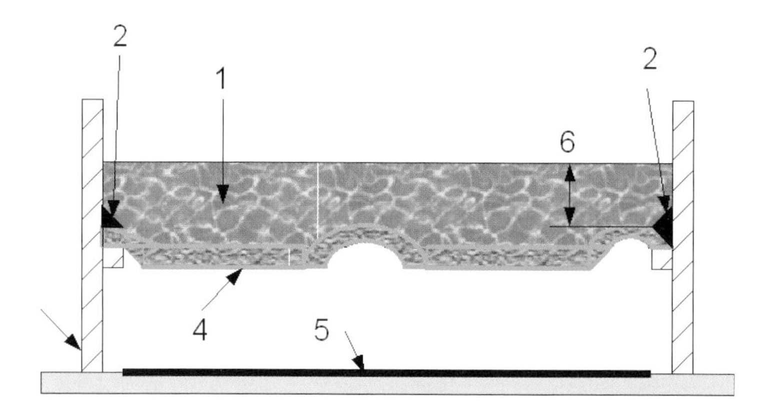

Fig. 1.56:
Water impermeability test method in accordance with DIN EN 491:2005-03 [1.74]
1 Water
2 Approx. 15 mm wide seal
3 Waterproof frame
4 Roof tile
5 Mirror
6 10 to 15 mm water layer

In the water impermeability test, water is applied to the roof tiles, which are then monitored for a certain period to detect any water penetrating through the tiles. For this purpose, a supporting frame is placed on top of the tile. A suitable mirror, located underneath the tile, is used to monitor droplet formation (Fig. 1.56).

Table 1.22: Specifications for concrete roof tiles and fittings in accordance with DIN EN 490:2006-09 [1.73], DIN EN 491:2005-03 [1.74] and DIN EN 13501-1 [1.113]

Product characteristic	Requirement	Test methods
Dimensions	Hanging length l_1 ± 4 mm Squareness: $l_2 - l_3 \leq 4$ mm Evenness: Permissible gap ≤ 3 mm The manufacturer must define dimensions, tolerances and measuring methods for fittings.	DIN EN 491
Mass	≤ 2 kg: ± 0.2 kg > 2 kg: ± 10%	DIN EN 491
Mechanical strength	F_{min} value not below the corresponding value specified in DIN EN 490	DIN EN 491 Structural strength
Water impermeability	Water droplets may occur on the underside, but must not detach before the end of the test period (20 hours).	DIN EN 491
Durability (freeze-thaw resistance)	Durability (freeze-thaw resistance) is deemed verified if the roof tiles comply with the requirements relating to water impermeability and strength.	DIN EN 491
Behaviour in the event of an external fire and flammability	Requirement is met if any existing external coating is inorganic or has a gross calorific value ≤ 4.0 MJ/m² or a mass of ≤ 200 g/m², or if the tiles conform to the provisions of Commission Decision 96/603/EC. Special provisions for flammability testing apply if the calorific value of the coating system exceeds a defined threshold.	DIN EN 13501-1

d) Cast stones
DIN V 18500:2006-12 [1.75] contains the specifications for the raw materials and manufacture of cast stones as well as the related product specifications and test methods. The individual product standards detail the specifications.

Cast stones made of concrete are subject to general specifications in the following areas:

- quality
- shapes and dimensions
- mechanical strength
- abrasive wear
- slip resistance
- weather resistance

A key characteristic of cast stones is their mechanical strength, which is determined by the tensile bending strength in the case of terrazzo tiles, steps and step coverings, and by the compressive strength for all other elements. Terrazzo tiles must conform to DIN EN 13748-2:2005-03 [1.68], whereas DIN EN 12390-5:2009-07 [1.85] applies to all other products in this category. This standard describes the tensile bending test with two-point load introduction as the reference method (Fig. 1.57). The alternative method specified in the standard, i.e. the three-point bending test (centred load application), may also be used; however, this must be clearly indicated in the test report. A key difference exists between centred loading and a two-point load application: the former method yields values that are 13% higher.

Cyclic freeze-thaw testing in accordance with DIN EN 13748-2:2005-03 [1.68] largely corresponds to the specification given in DIN EN 1338:2003-08 [1.65].

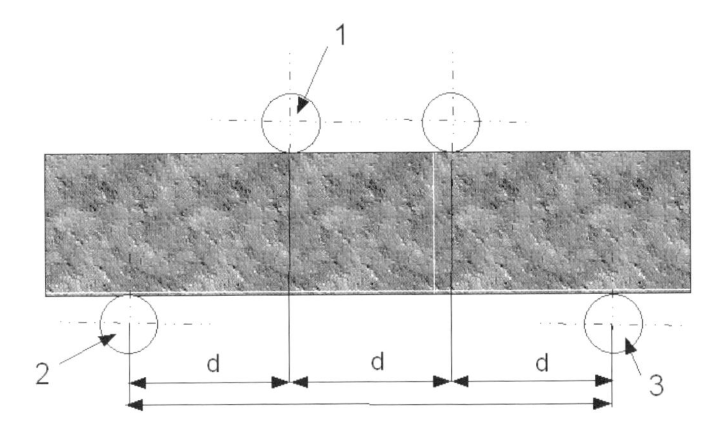

Fig. 1.57:
Principle of the tensile bending test with two-point load introduction in accordance with DIN EN 12390-5:2009-07 [1.85]
1 Load roller (tilt-and-turn design)
2 Support roller
3 Support roller (tilt-and-turn design)

Table 1.23: Specifications for concrete cast stones conforming to DIN V 18500:2006-12 [1.75]

Product characteristic	Requirement	Test methods
Dimensional limits and evenness tolerances	Depending on the product groups and dimensions of the products	Measurement
Surface quality	Projections, depressions, cracks or spalling not permitted	Visual inspection from a distance of 2 m
Face concrete thickness	Tiles: ≥ 8 mm Steps, step coverings: ≥ 15 mm	
Concrete cover	Specifications of DIN 1045-1	DIN 1045-1 [1.46]
Weather resistance	≤ 7% (outdoor use) DIN EN 13748-2 applies to terrazzo tiles	DIN EN 13748-2 [1.68]
Abrasive wear	Hardness class I or 2[1]: 18 $cm^3/50\ cm^2$ Hardness class II or 3[1]: 20 $cm^3/50\ cm^2$ Hardness class III or 4[1]: 26 $cm^3/50\ cm^2$	DIN 52108 [1.123]
Tensile bending strength	For floor tiles, steps and step coverings: ≥ 5.0 N/mm^2 ≥ 4.0 N/mm^2 (single value)	DIN EN 12390-5 using test specimens [1.85] with specified dimensions
Compressive strength	Class C25/30 as specified in DIN EN 206-1 [1.18] applies to all other items	DIN EN 12504-1 using core samples [1.88] DIN EN 12390-3 [1.83] using specially produced test specimens
Slip resistance	Items ground with a 220 grit have a sufficient slip resistance if they are not polished	If required, according to DIN EN 13748-2 [1.68] following prior agreement

[1] in accordance with DIN EN 13748-2

Table 1.23 summarises the specifications and test methods applied.

1.3.2.2 Requirements for precast elements

Apart from specifications for building materials and production, DIN EN 13369:2004-09 [1.76] also contains the basic requirements for the finished product and the test methods to be used. The individual product standards list the detailed specifications.

1.3.2.3 Requirements for concrete pipes and manholes

Having been in force since 24 November 2004, DIN EN 1916:2003-04 [1.77] and DIN V 1201:2004-08 [1.78] pertaining to concrete and reinforced concrete pipes must always be applied together in order to maintain the national safety level.

A distinction must be made between type 1 and type 2 concrete pipes with respect to concrete compressive strength. In general, the concrete must meet the specifications stated in DIN EN 206-1:2001-07 [1.18] and DIN 1045-2:2008-08 [1.17]. For type 1 pipes, the concrete must conform to exposure class XA1 (environment with minor chemical attack) whereas type 2 pipes must meet the requirements of class XA2 (moderate chemical attack) or XM2 (strong wear).

Table 1.24: Specifications for precast concrete elements in accordance with DIN EN 13369:2004-09 [1.76]

Product characteristic	Requirement	Test methods
Dimensions and surface quality Dimensional limits	Specified in the relevant product standards $\Delta L = \pm (10 + L/1000) \leq \pm 40$ mm	DIN EN 13369, Annex L Measurement and visual inspection
Potential concrete strength	Compressive strength for the proof of concrete quality is derived from the potential strength	DIN EN 12390-2 [1.82] or DIN EN 12390-3 [1.83] Cube or cylinder Age of 28 days at time of testing
Component strength		DIN EN 12504-1 [1.88] using core samples taken from the component
Water absorption	Must be specified if required in the relevant product standard	DIN EN 13369 Annex J
Dry bulk density	Must be specified if required in the relevant product standard	DIN EN 12390-7 [1.87]
Finished product	Reference tests are described in the relevant product standards	
Weight of precast element	± 3% Must be specified if required in the relevant product standard	
Fire rating and fire behaviour	Must be specified if required in the relevant product standard (cement-bound precast concrete elements, Class A.1; test not required)	
Soundproofing characteristics (airborne and impact sound insulation)	Must be specified if required in the relevant product standard	DIN EN ISO 140-3 [1.89] DIN EN ISO 140-6 [1.90]
Thermal insulation characteristics	Must be specified if required in the relevant product standard	DIN EN 12664 [1.91] Thermal conductivity of the building material, or DIN EN ISO 6946 [1.92] or DIN EN ISO 8990 [1.93] or DIN EN 1934 [1.94] Thermal conduction resistance of the component

Type 1 pipes must exhibit a concrete compressive strength that corresponds to class C35/45, whereas type 2 pipes must conform to class C40/50 as specified in DIN EN 206-1:2001-07 [1.18] and DIN 1045-2:2008-08 [1.17]. Type 1 and 2 jacking pipes must at least conform to compressive strength class C40/50.

The main specifications for concrete pipes relate to:

- quality
- shapes and dimensions
- water absorption
- crushing strength
- longitudinal bending strength and
- water impermeability

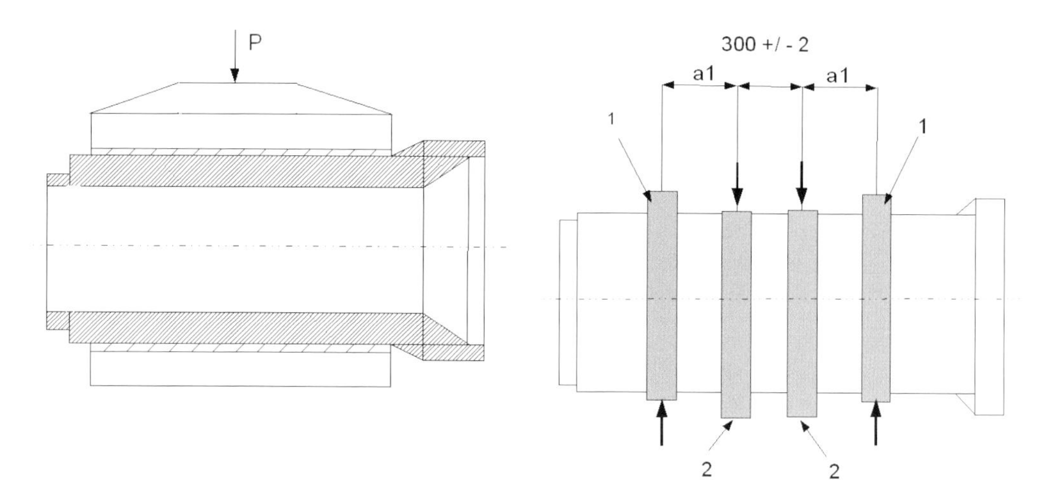

Fig. 1.58:
Permissible set-ups for crushing strength tests of circular pipes (excluding pipes with DN ≤ 1,200) in accordance with DIN EN 1916:2003-04 [1.77]

Fig. 1.59:
Arrangement of loading and support for the four-point loading test in accordance with DIN EN 1916:2003-04 [1.77]
1 Support strap
2 Load strap
The lever arm a1 must be greater or equal to 300 mm

Concrete pipes are mainly characterised by their crushing strength, longitudinal bending strength and water tightness.

For the purpose of testing the crushing strength of pipes as specified in DIN EN 1916:2003-04 [1.77], the test set-up must consist of a machine that is capable of applying the full test load (P) without any impact or shock and within a tolerance range of 3% from the specified load. This involves inserting the concrete pipe into the test set-up as shown in Fig. 1.58. The same applies to pipes with bases. The type of load to be applied according to the standard is governed by the material used for the pipe (i.e. non-reinforced, reinforced or steel-fibre reinforced concrete). The pipe may be kept wet for a period of up to 28 hours prior to the test.

The longitudinal bending strength must be measured for circular pipes ≤ DN 250 with lengths greater than six times their external diameter. The test may be carried out on a section of a circular pipe (with or without a socket) with a length greater than 1.25 m or on an entire circular pipe. It is up to the producer to decide whether to keep the test specimen wet for a period of up to 28 hours prior to the test.

The main test is the four-point loading method (Fig. 1.59). The load must be applied to the specimen without vibration or impact, and increased in uniform increments of 6 kN to 9 kN per minute.

Individual components or pipe connections must not leak or exhibit other visible defects during the testing period. The water impermeability test is thus carried out to assess whether elbowed pipe connections and/or connections in shear and structural components pass a hydrostatic test, i.e. whether they remain watertight when the specified test pressure is applied. For this purpose, the items must be firmly restrained in the test rig. Pipe ends must be sealed by appropriate means, and the specified test pressure must be applied for the required period (15 minutes) after the pipes have been filled with water. The pressure must not exceed the specified level by more than 10% and must not fall below this level.

The hydrostatic test of individual items is used to assess the tightness of the concrete. No such test is required for products with a wall thickness exceeding 125 mm.

When testing the pipe union, the test set-up must be designed to accommodate two pipes that are connected to each other complete with their seals. They must be flexibly connected and positioned in such a way that they can move toward each other until they reach the set limits. The pipe connections may be tested either in an elbow arrangement or in shear, or in a test combining these two characteristics.

In the first case, both pipes are cautiously bent, filled with water and subjected to an internal test pressure of 50 kPa. This pressure level must be kept constant for fifteen minutes. In the second arrangement, the set-up shown in Fig. 1.60 is used. The pipes are filled with water and de-aerated. Both the test pressure of 50 kPa and the shear force are applied and kept constant for fifteen minutes. The pipe connection is tested for conformity.

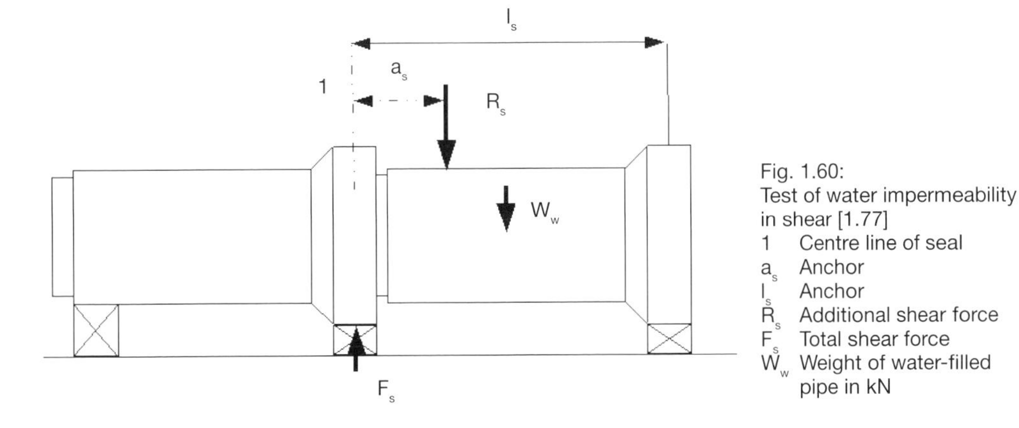

Fig. 1.60:
Test of water impermeability in shear [1.77]
1 Centre line of seal
a_s Anchor
l_s Anchor
R_s Additional shear force
F_s Total shear force
W_w Weight of water-filled pipe in kN

Table 1.25: Testing requirements for pipes and fittings in accordance with DIN EN 1916: 2003-04 and DIN V 1201: 2004-08 [1.77], [1.78]

Product characteristic	Requirement	Test methods
Dimensions	Internal diameter tolerances depend on the nominal bore of the pipes and must lie between ± 2 and ± 15 mm. For each produced type 2 concrete, reinforced or steel-fibre reinforced concrete pipe with a nominal bore of up to and including DN 1,000, the external diameter of the spigot end d_{sp} must be measured.	Measurement
Visual inspection of surface quality	Uniform, closed condition, preventing any deleterious effect on serviceability and hydraulic performance. Max. pore diameter 15 mm, depth of 10 mm permitted; in reinforced concrete pipes, minimum concrete cover of 10 mm above pores. Spiderweb-like hairline cracks with a width ≤ 0.15 mm are permissible.	Visual inspection
Water absorption	Water absorption ≤ 6 M.-%	DIN EN 1916, Annex F using test specimens
Crushing strength	Grouping in load classes. Resistance to the minimum crushing force depending on the nominal bore and strength class; the minimum crushing force is the short-term test force. It is equivalent to the product of the strength class and the nominal bore/1,000. For reinforced and steel-fibre reinforced concrete pipes, the cracking force F_c must also be verified.	DIN EN 1916, Annex C
Longitudinal bending strength	During the test, the longitudinal bending moment M of a pipe must not be lower than the moment calculated using the following equation: $M = C \cdot DN \cdot l^2$ (kNm) C – constant: 0.013 (kN/m) DN – nominal bore l – length (m)	DIN EN 1916, Annex D
Water impermeability	The following applies to wall thicknesses < 125 mm: each produced concrete, reinforced or steel-fibre reinforced concrete pipe of type 2 with a nominal bore of up to 1,000 must remain watertight during a short-term factory test with a positive water pressure of 1 bar, a positive air pressure of up to 20 kPa (0.2 bar) or a negative air pressure of 20 kPa (0.2 bar).	DIN EN 1916, Annex E Tests to be carried out for individual components or two components connected to each other
Reinforcement and concrete cover	Determination of concrete covers depending on the ambient conditions in accordance with DIN V 1201; the minimum concrete cover of external surfaces of jacking pipes must be increased by 5 mm according to DIN EN 1916. Ring and longitudinal reinforcements and concrete covers must be checked for conformity with the factory documents (special rules for jacking pipes)	DIN EN 1916 Testing of a reinforced concrete pipe segment. Exposure of the reinforcement, measurement of concrete cover, documentation of smallest dimension rounded to nearest millimetre.
Core sample strength (only applicable to jacking pipes)	Concrete compressive strength class C40/50 Concrete compressive strength ≥ 40 N/mm²	DIN 1048-2 [1.96] Tests of core samples taken at intervals of one third of the pipe length.

Table 1.25 lists the specifications for finished products and the test methods to be used.

Precast concrete manhole elements according to DIN V 4034-1:2004-08 [1.80] and DIN EN 1917:2003-04 [1.79] are governed by specifications similar to those defined for pipes and fittings made of non-reinforced, reinforced and steel-fibre reinforced concrete. The two standards must be applied together.

Type 1 precast manhole elements must also conform to compressive strength class C35/45. These items are mainly used for storm drains.

Type 2 precast manhole elements comply with the specifications of exposure class XA2, as well as XM2 if required, in accordance with DIN EN 206-1:2001-07. They correspond to the quality standard previously applied in Germany and are particularly suitable for combined sewers and foul water sewers. Type 2 elements must conform to compressive strength class C40/50 (see DIN EN 206-1:2001-07 and DIN 1045-2:2008-08). This specification also applies to drains and footholds of manhole bases that are produced in a single cast together with the floor and the shaft. The concrete for subsequently installed and lined drains and treads must have the same compressive strength as the concrete used for the manhole elements. In the hardened state, it must achieve a compressive strength that at least corresponds to strength class C16/20 (drain concrete).

Precast manhole elements made of concrete and reinforced concrete must be interchangeable, provided that the same types of connections and climbing systems are used.

General specifications for manhole elements made of precast concrete relate to their:

- quality
- shapes and dimensions
- water absorption
- crushing strength
- vertical strength and
- water impermeability

The horizontal or vertical arrangement shown in Fig. 1.61 and Fig. 1.62 may be used to test the crushing strength of manhole units.

The vertical strength test applies to transition parts and covers. The test rig must consist of steel or cast iron plates that apply the required test load to the element depending on its position. Support widths for the access or inspection shaft must be the same as in the installed condition. The minimum vertical test force must be applied above the opening, as shown in Fig. 1.63 and Fig. 1.64. In this process, the load should be increased to failure in a smooth and uniform manner.

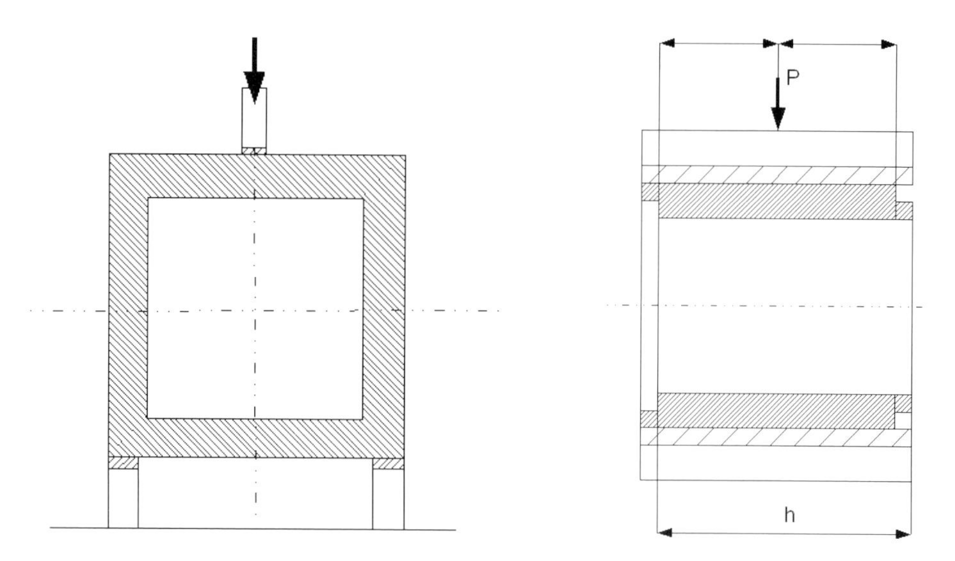

Fig. 1.61: Crushing strength test for components in horizontal position in accordance with DIN EN 1917:2003-04 [1.79]

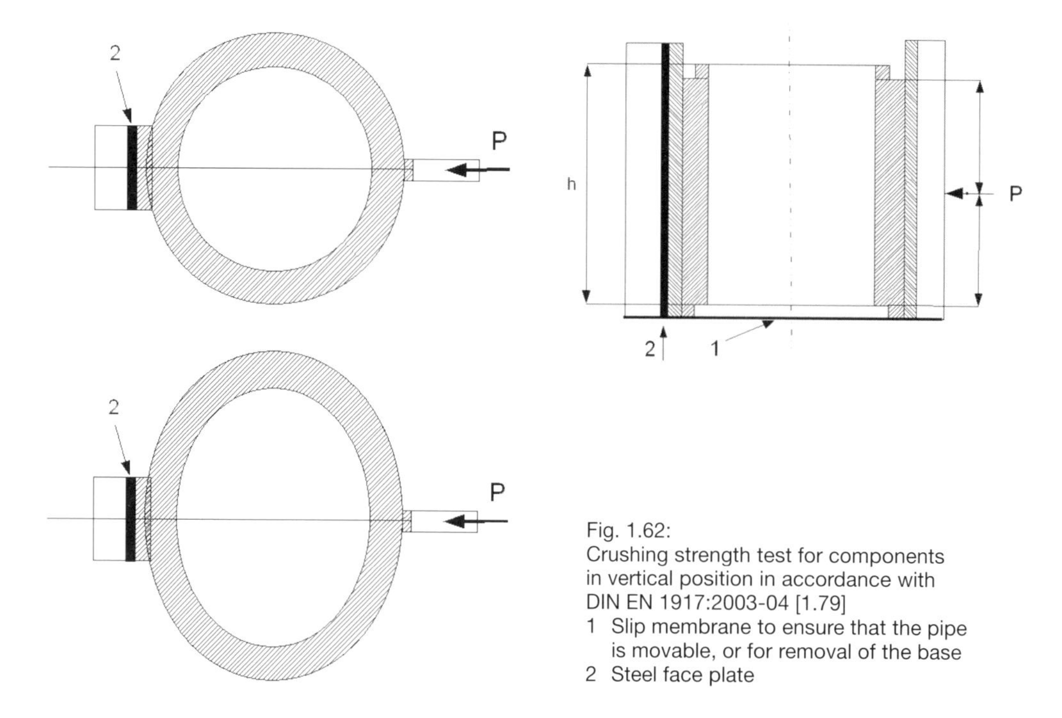

Fig. 1.62:
Crushing strength test for components in vertical position in accordance with DIN EN 1917:2003-04 [1.79]

1 Slip membrane to ensure that the pipe is movable, or for removal of the base
2 Steel face plate

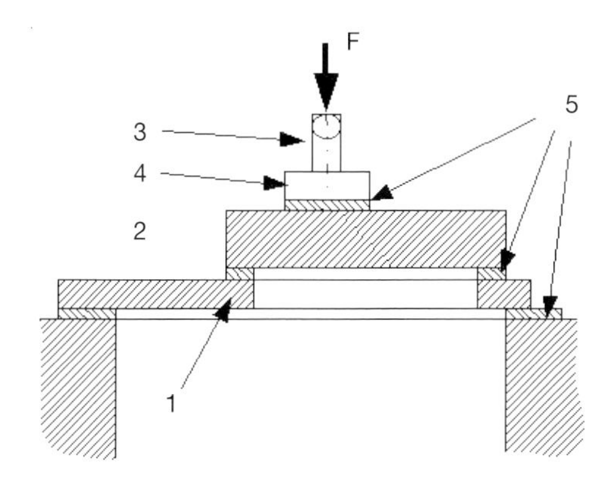

Fig. 1.63: Test of the vertical strength of cover components in accordance with DIN EN 1917:2003-04 [1.80]
1 Slab
2 Steel or cast iron plate
3 Ball-and-socket joint
4 Loading plate 300 mm x 300 mm
5 Support

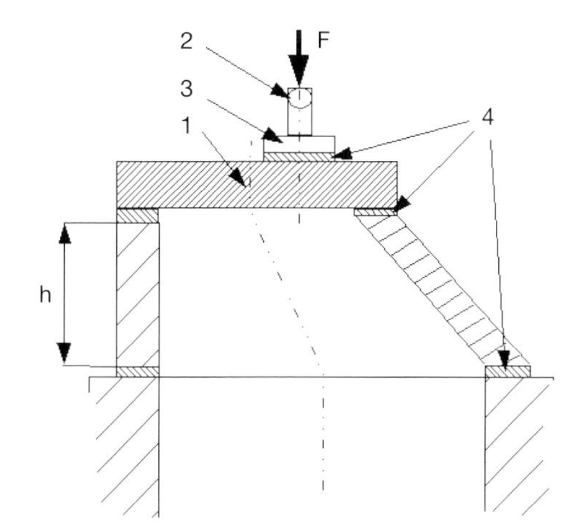

Fig. 1.64: Test of the vertical crushing strength of cones [1.80]
1 Steel or cast iron plate
2 Ball-and-socket joint
3 Loading plate 300 mm x 300 mm
4 Rubber or gypsum, thickness 20 mm ± 5 mm

Table 1.26: Test specifications for precast manhole units conforming to DIN EN 1917:2003-04 and DIN V 4034-1: 2004-08 [1.80], [1.81]

Product characteristic	Requirement	Test methods
Dimensions	Internal diameter tolerances depend on the nominal bore of the pipes and must lie between ± 8 and ± 10 mm. The evenness tolerance of the spigot end is 5 mm for manhole rings. For reasons of interchangeability, the d_{sp}, l_{sp}, l_{so} and l_s dimensions are specified in DIN V 4034-1.	Measurement
Visual inspection of surface quality	Sealing surfaces of the connecting sections must be free of irregularities. Spiderweb-like hairline cracks with a width ≤ 0.15 mm are permissible.	Visual inspection
Water absorption	Water absorption ≤ 6 M.-%	DIN EN 1917, Annex D using specimens
Crushing strength	The minimum crushing strength F_n must correspond to the nominal bore and strength class of the precast manhole unit. For manhole rings, the minimum crushing strength F_n is 80 kN/m. Reinforced concrete manhole rings must additionally resist a test force F_c amounting to $0.67 \cdot F_n$ (maximum crack width in the concrete tensile zone after relief 0.3 mm). (see also DBV Merkblatt [Code of Practice] on steel-fibre reinforced concrete for special requirements for steel-fibre reinforced concrete)	DIN EN 1917, Annex A Choice of test set-up depends on the cross-sectional shape

Table 1.26: Test specifications for precast manhole units conforming to DIN EN 1917:2003-04 and DIN V 4034-1: 2004-08 [1.80], [1.81] *(continued)*

Product characteristic	Requirement	Test methods
Vertical strength	The vertical minimum crushing strength F_v of cover slabs, transitions, cover components and cones must be 300 kN. The vertical test force F_p must reach 120 kN (maximum crack width in the concrete tensile tone after relief: 0.15 mm)	DIN EN 1917, Annex B
Water impermeability	The following applies to wall thicknesses < 125 mm: type 2 precast manhole units made of concrete or reinforced concrete must remain watertight at an internal positive pressure of up to 1.0 bar. During the 15-minute testing period, the amount of added water must not exceed 0.07 l per m² of wetted surface. Moist spots on the outside surface do not constitute defects.	DIN EN 1917, Annex C Test of two precast manhole components linked to each other by a single connection (internal quality control at the factory).
Concrete cover	Minimum concrete cover c_{min} = 25 mm. The nominal cover c_{nom} must be defined in the factory documentation.	DIN EN 1917 Test of a reinforced concrete pipe segment, exposure of the reinforcement, measurement of concrete cover, documentation of smallest dimension rounded to nearest millimetre.
Reinforcement	Must correspond to the factory documentation.	DIN EN 1917 The position and amount of ring reinforcement must be tested over a length of at least 1 m or over the entire height of the component. Check conformity with the factory documents.
Concrete compressive strength (for manhole bases, walls of cover components, alignment pieces and cones)	Concrete compressive strength class C40/50	Test specimen according to DIN 1048-5 [1.95] (internal quality control) Tests of core samples according to DIN 1048-2 [1.96] taken at intervals of one third of the component height.

Table 1.26 describes the testing specifications for precast manhole units.

The step irons installed in manholes must also be tested for vertical loading and horizontal pull-out force (Annex E to DIN EN 1917:2003-04 [1.79]).

1.3.3 Evaluation of Conformity

1.3.3.1 Fundamentals

An internal quality control system must be in place for evaluating the conformity on the basis of the European standard. This system includes quality control during the production process at the factory. This internal system ensures that produced items comply with the defined specifications.

It is a documentation-based system that comprises both the preparation of processes, including related instructions, and their effective implementation. The tests required for the proof of conformity must be carried out using the methods specified in the European standards, and must be recorded in an appropriate manner.

The factory logbook must include the following information:

- description of the products
- date of manufacture
- test methods
- test results
- applied dimensional tolerances
- signature(s) of the employee(s) performing the internal quality control

The logbook must be retained for a period of five years.

Inspection and testing schedules should be used for periodic testing and inspections of the individual materials, plant and equipment, production processes and laboratory facilities. Such schedules must comply with the applicable minimum frequencies. The European standard includes special testing schedules for finished products, which should be subject to a particularly thorough quality control.

The involvement of a certifying body (i.e. a recognised quality control association or officially recognised testing institution commissioned on the basis of a quality control agreement) serves to evaluate the products and to perform initial tests to check whether the specifications of the relevant European standard are adhered to and whether the requirements for the production process and for the agreed factory control system are met. A monitoring programme is implemented to periodically check compliance with relevant specifications. The outcomes of periodic external quality control activities must be documented in test reports.

The European standards describe procedures to be followed for issuing a certificate of conformity for the CE marking of the products. The manufacturer must issue a declaration of conformity in the case of compliance with the specifications and conditions of the relevant standard. This declaration makes it possible for the manufacturer to CE-mark its products.

1.3.3.2 Conformity of small concrete products

a) Concrete products for road construction
The manufacturer must prove conformity of the product with the specifications of DIN EN 1338:2003-08 [1.65], DIN EN 1339:2003-08 [1.66] or DIN EN 1340:2003-08 [1.67], as well as with the values (ranges or classes) stated with regard to product properties. To do so, the manufacturer must take the following steps:

- a type test of the product and
- internal quality control measures at the factory, including a product check.

In addition, the conformity of the product with this standard can be evaluated either by an external quality control scheme, within which the type test performed by the manufacturer and the processes of internal quality control are monitored, or by an acceptance test carried out for the respective delivery upon product handover.

Type tests comprise initial and subsequent type tests. An initial type test is performed when the manufacture of a new product type or range commences, or when a new production line is to be commissioned. Subsequent type tests of the relevant characteristic must be carried out when a change in the raw materials used, the mix design or the production equipment or process may result in significant changes to some or all of the characteristics of the finished product.

Regular type tests must be performed for abrasion and weather resistance, even if no modifications were implemented.

The internal quality control system comprises the procedure as well as the regular checks and tests that must be carried out at the manufacturer's factory. A sampling and testing plan must be prepared and implemented for the testing of products. The test results must meet the defined conformity criteria.

For the purpose of CE-marking concrete paving blocks, flags and kerbstones, System 4 of the certificate of conformity (Directive 89/106/EEC (CPD), Annex III, 2 (ii), third possibility) must be applied, which comprises an initial test and internal quality control measures at the factory. Key parameters are durability and tensile or tensile bending strength, as well as sliding and slip resistance and thermal conductivity, if applicable.

b) Concrete masonry units
The manufacturer must prove conformity of the product with the specifications of DIN EN 771-3:2005-05 [1.97], and with the declared values that describe the product properties, by:

- an initial product test and
- internal quality control measures at the factory.

When a new product has been developed, and prior to its market launch, appropriate initial tests must be carried out to ensure that the actual characteristics of the product meet the specifications contained in this European standard, and that the product-specific values stated by the manufacturer are adhered to. The initial test must be repeated if material changes to the mix design are made.

The internal quality control system comprises the identification and monitoring of the raw materials, the production process, and testing of finished products and stocks.

The CE marking of concrete masonry units is based on DIN EN 771-3:2005-05 [1.97], which specifies categories I and II for these products. For category I products, System 2+ in accordance with Directive 89/106/EEC (CPD), Annex III, 2 (ii), possibility 1, must be used, including certification of the internal quality control system by a notified body on the basis of an initial inspection of the factory and the internal quality control system. In addition, the internal quality control system must be monitored, evaluated and verified on an ongoing basis. Key parameters relevant to concrete masonry units are compressive strength, dimensional stability and bond strength. System 4 applies to category II products.

c) Concrete roof tiles
The conformity of concrete roof tiles and fittings with the specifications of DIN EN 490:2006-09 [1.73] must be proven by:

- initial testing and
- internal quality control measures at the factory

Roof tiles with common characteristics may be grouped in product families.

The following provisions apply to products used in the European Economic Area:

The choice of the system to be used to prove conformity is governed by the product's intended use. All concrete roof tiles and fittings to be tested are subject to System 3 of the attestation of conformity (see EU Directive 89/106/EEC (CPD), Annex III, 2 (ii), possibility 2). Under System 3, the manufacturer must implement internal quality control measures and its own initial testing, as well as initial testing by a notified body.

Products that do not require testing for flammability are tested according to System 3 of the certificate of conformity.

d) Cast stones
The conformity of cast stone items with the specifications of DIN V 18500:2006-12 [1.75] must be proven by:

- initial testing
- the internal quality control system implemented by the manufacturer
- external quality control and certification

DIN 18200:2000-05 [1.121] governs the procedure to be followed for the attestation of conformity. Structural and stiffening cast stone elements are subject to the provisions of DIN EN 206-1:2001-07 [1.18] in conjunction with DIN 1045-2: 2008-08 [1.17] and DIN 1045-4:2001-07 [1.122] as regards their initial testing, internal and external quality control and certification.

An initial test must be carried out to prove the conformity with the specifications of DIN V 18500:2006-12. The test specimens to be used for this purpose must be at least 28 days old. The initial test must be carried out by the external quality control institution.

The internal quality control system must include the procedures, the regular checks and tests and use of the results for the management and modification of raw materials and of other materials used for the manufacture of the products, as well as of the equipment, production process and product properties.

External quality control and certification must be carried out by an industry-wide quality control association or by an appropriate quality control and certification entity on the basis of a contract.

1.3.3.3 Conformity of precast elements

The conformity of the product with the specifications of DIN EN 13369: 2004-09 [1.76] and with the defined or declared values (ranges or classes) pertaining to the product properties must be proven by:

- an initial product test
- internal quality control at the factory, including product testing

In addition to these requirements, the conformity of the product may be evaluated by an accredited body. The accredited body may assess the conformity of the internal quality control system with respect to the following individual tasks:

- initial inspection of the factory and of the internal quality control system
- ongoing monitoring, evaluation and confirmation of conformity of the internal quality control system

The accredited body may also evaluate product conformity on the basis of the following tasks:

- monitoring, evaluation and confirmation of conformity of the initial product test
- random testing of samples taken at the factory or, if required, on the construction site

The tasks and responsibilities of the accredited body depend on the specific product.

The initial inspection of the factory and the internal quality control system serves to prove conformity with the specifications contained in the standard.

The internal quality control system must include the relevant procedures, instructions, regular checks and tests, as well as the availability of the results for the testing of the equipment, raw materials and other supplied materials, the production process and the product. The accredited body reviews the conformity with the relevant requirements and the existence of a testing schedule.
A representative of the accredited body must attend the initial product tests or carry out these tests him- or herself.

The reliability of the internal quality control results should be evaluated by taking random samples and testing them according to a related schedule.

The system to be used for the certificate of conformity for CE marking purposes depends on the specific applications of the precast concrete elements. For load-bearing elements, System 2+ (Directive 89/106/EEC (CPD), Annex III, -2 (ii), possibility 1) should be mainly used, including certification of the internal quality control system by a notified body on the basis of an initial inspection of the factory and the internal quality control system. In addition, the internal quality control system must be monitored, evaluated and verified on an ongoing basis.

1.3.3.4 Conformity of pipes and manholes

Type 2 pipes and precast manhole elements designed for an environment with a "moderate chemical attack" must be produced, tested and quality-controlled in accordance with DIN EN 1916:2003-04 [1.77] and DIN V 1201:2004-08 [1.78], or DIN EN 1917:2003-04 [1.79] and DIN V 4034-1:2004-08 [1.80], in order to achieve the safety parameters required in Germany.

To evaluate conformity, the specified characteristics must be ascertained by a suitability test (initial test) and ensured by a quality control system that comprises an internal and external monitoring component.

An initial test is necessary when the production of a new type commences, or if material changes are made to the design, materials or production process.

The internal quality control system includes the monitoring of the product properties at the factory. DIN V 1201:2004-08 [1.78] and DIN V 4034-1:2004-08 [1.80] define the type, scope and intervals at which the required tests must be carried out.

The external part of the system comprises the control of the results of internal quality control measures. In addition, product properties are checked on a random basis. The certifying body may issue a certificate to the manufacturer after a successful initial inspection of the production facility and a positive evaluation of the external quality control component.

For the purpose of CE-marking of pipes, fittings and precast manhole units, System 4 of the certificate of conformity (Directive 89/106/EEC (CPD), Annex III, 2 (ii), third possibility) must be applied, which comprises an initial test and internal quality control measures at the factory. The key parameters to be checked for pipes and fittings are dimensional tolerances, crushing strength, longitudinal bending strength, water impermeability and durability. For manhole units, these include the checking of inspection openings, mechanical strength, structural strength of installed step irons, water impermeability and durability.

1.4 Fundamentals of Plant and Equipment

1.4.1 Vibration Exciter Systems

All of the various types of equipment used for the manufacture of concrete products include typical, repetitive components. This applies particularly to vibration exciters, which are mainly used for the moulding and compaction of concrete products.

Various physical principles are used to generate the required loads. Vibration exciters can generally be classified using the criteria shown in Fig. 1.65.

Concrete is compacted primarily by vibrators with unbalance exciters, which exhibit high performance levels and a simple design. Section 1.1.4.3 describes their mechanisms of action and the calculation of the centrifugal force. Both internal and external vibrators are used.

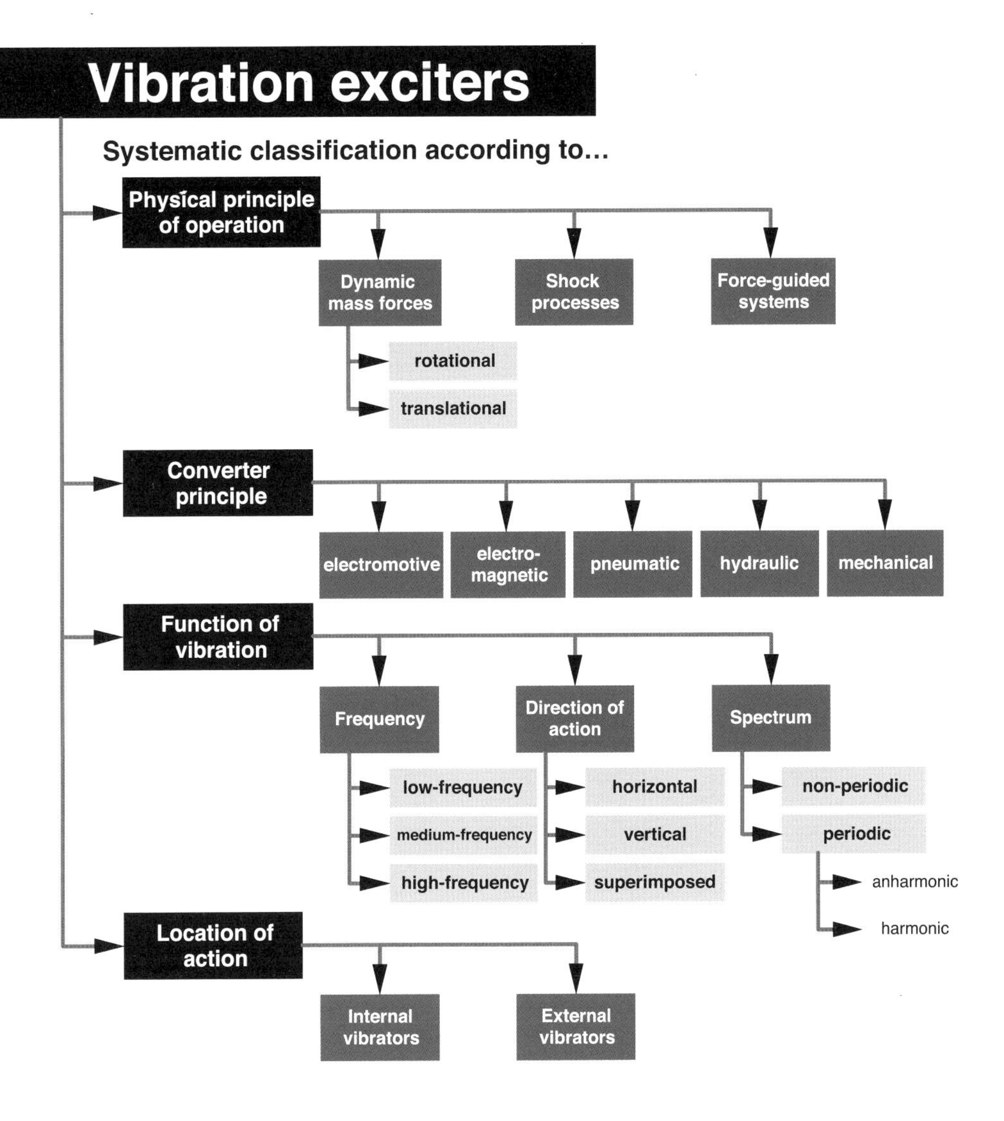

Fig. 1.65: Classification of vibration exciter systems

Universal motors (universal current and collector motors) and asynchronous motors (three-phase current motors) can be used to drive electric vibrators [1.120]. High-performance electric vibrators for the compaction of concrete are mainly equipped with asynchronous motors, which are significantly more robust than universal motors and require a lower amount of maintenance. Fig. 1.66 shows a section of an electric external vibrator fitted with an asynchronous motor.

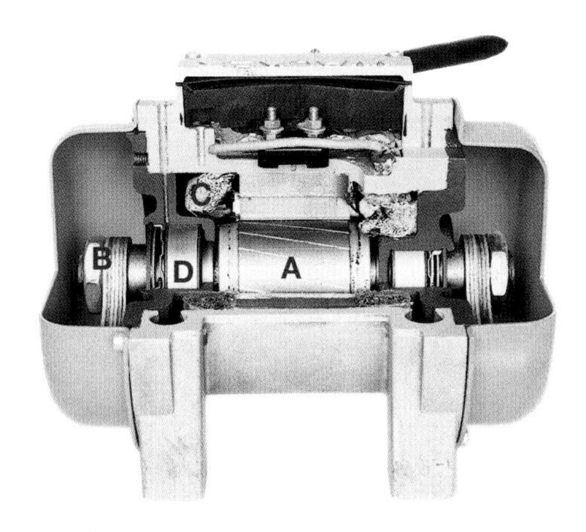

Fig. 1.66:
Cut-out view of an electric external vibrator equipped with an asynchronous motor. The figure shows: the rotor (A) with unbalances (B), electrical winding of the stator (C), ball bearings (D) and electrical connections (E)

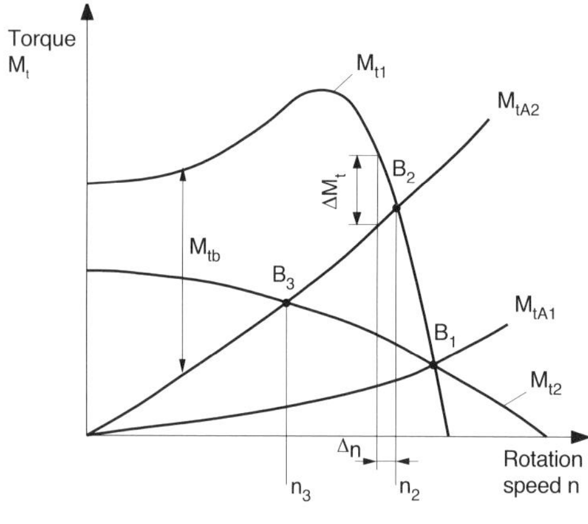

Fig. 1.67:
Performance curves

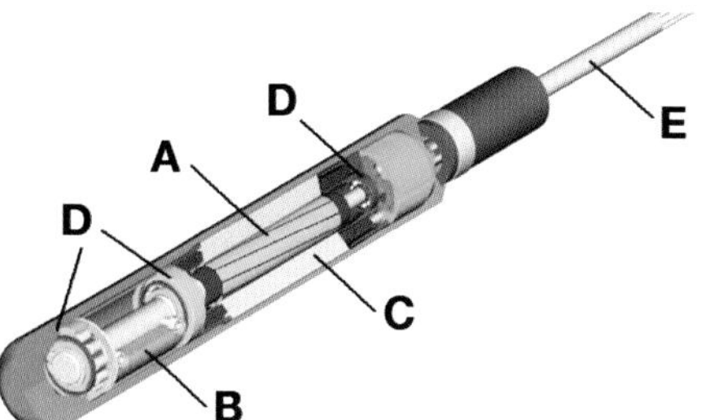

Fig. 1.68:
Design of a high-frequency internal vibrator by WACKER: Drive shaft with rotor (A) and unbalance (B), stator unit (C), roller bearing (D) and electrical supply lines (E)

Fig. 1.67 includes a qualitative representation of the performance curves M_t (n) for an asynchronous motor (M_{t1}) and a universal motor (M_{t2}). The operating points B are represented by the intersections with the assumed load curves M_{tA}. The diagram shows that the speed of the asynchronous motor (n_2) decreases to a significantly lesser extent than that of the universal motor (n_3) when the moment of the machine, $M_{tA1} \rightarrow M_{tA2}$, is increased.

The higher accelerating moment M_{tb} of the asynchronous motor also results in a shorter acceleration time t_b to the operating point. Due to the large torque differences ΔM_t

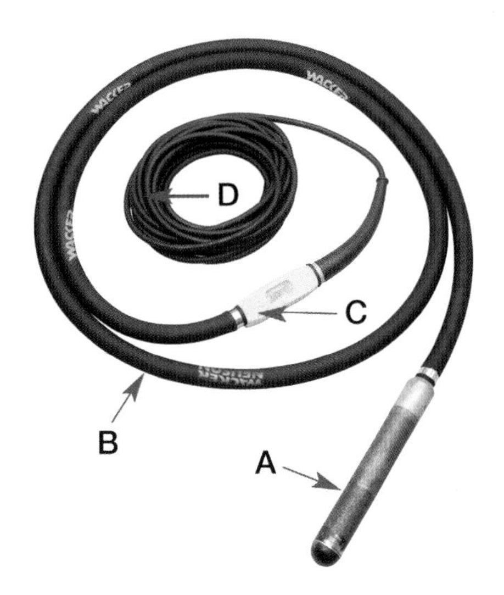

Fig. 1.69:
WACKER IREN 57 high-frequency internal vibrator with vibrating cylinder (A), protective hose (B), switch (C) and electrical supply line (D)

between M_{t1} and M_{tA1} or M_{tA2}, the asynchronous motor exhibits a very stable system performance at the operating point because the condition is met:

$$\frac{dM_t}{dn} < \frac{dM_{tA}}{dn} \tag{1.31}$$

Internal vibrators consist of a vibrating cylinder within which an unbalance mass rotates (Fig. 1.68).

In most cases, the cylinder is attached to a flexible hose or shaft that is used to guide the cylinder movement during the immersion phase (Fig. 1.69).

1.4.2 Research and Development

Several phases of research and development of technological processes and equipment for the manufacture of concrete products can be distinguished:

1. Modelling and simulation of the workability behaviour of mixes
2. Verification of the results of modelling investigations on a laboratory scale
3. Dynamic modelling and simulation of production equipment
4. Verification of dynamic modelling results on a pilot scale
5. Development of quality assurance systems
6. Implementation of research and development results to practical conditions and their subsequent review; measurement verification.

1.4.2.1 Modelling and simulation of the workability behaviour of mixes

It is crucial to acquire an in-depth knowledge of the workability characteristics of mixes and their mathematical description for the design of machines, the implementation of new processing methods and the development of new building materials systems. The more accurately the related models describe the processing sequence, the better a process can be simulated and the required equipment designed, and the lower the likelihood of production errors or defects.

a) Processing sequences

The processing sequence of a concrete mix comprises all process steps – from batching and mixing to transport and pouring, and finally compaction and demoulding (Fig. 1.70). For all sub-processes, key parameters describe the process and its outcome [8].

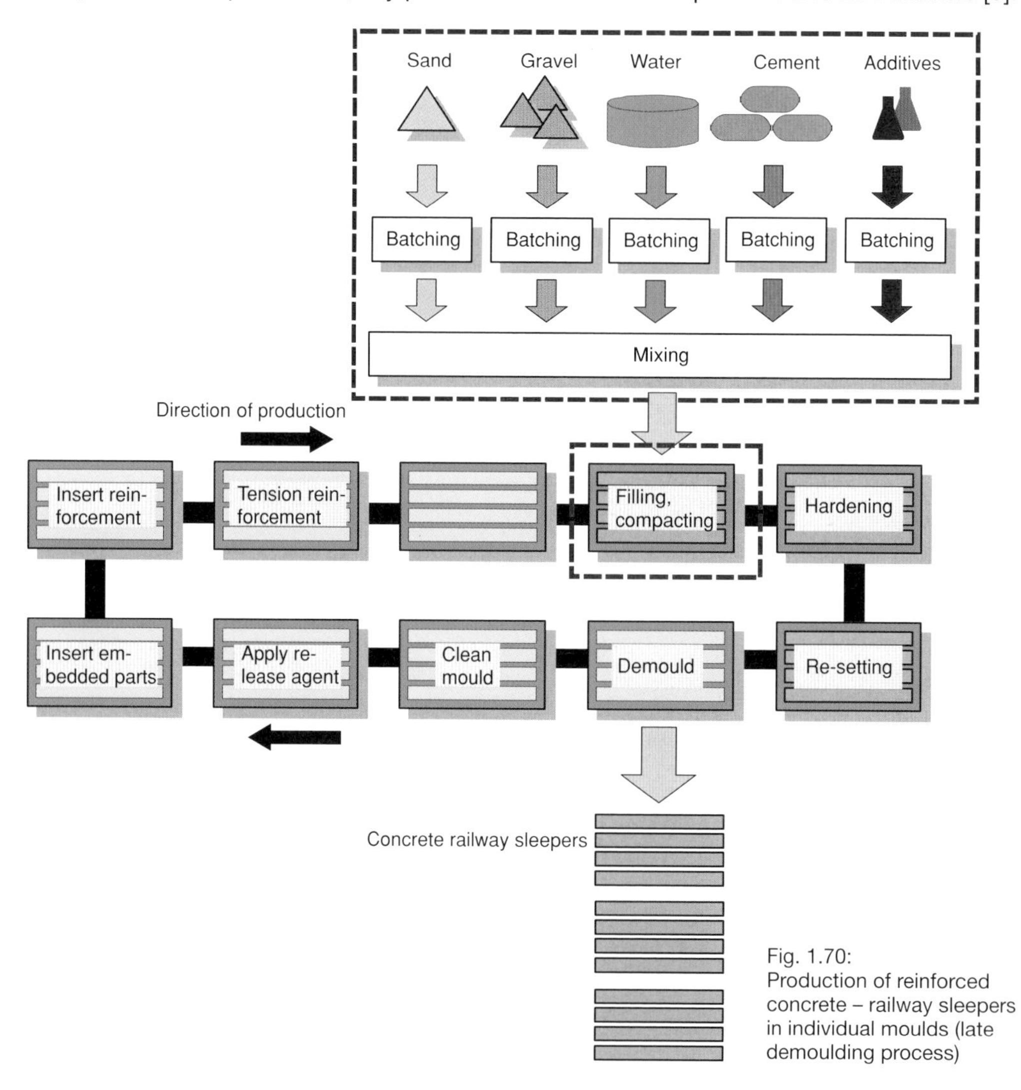

Fig. 1.70: Production of reinforced concrete – railway sleepers in individual moulds (late demoulding process)

In the production process, material- and process-related as well as equipment- and product-specific aspects need to be mutually harmonised whilst taking account of the fact that the workability characteristics of the concrete mix have a significant influence on the individual processes.

b) Simulation methods
Depending on the specific task to be performed, the processing behaviour of mixes can be analysed by applying structural mechanics, fluid mechanics or corpuscular

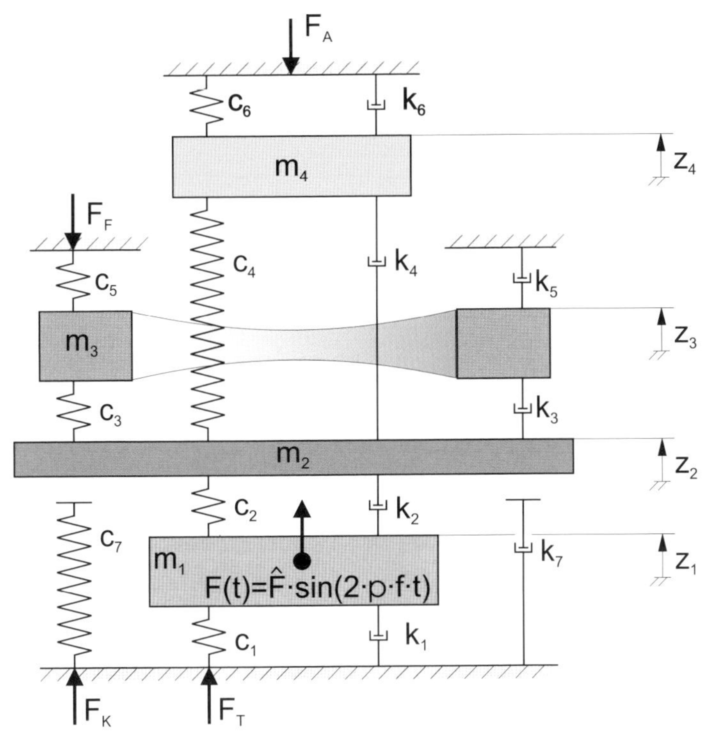

m_1	Mass of table
m_2	Mass of baseboard + ½ concrete mass
m_3	Mass of the lower mould section
m_4	Mass of the upper mould section and loading plate + ½ concrete mass
$z_1 \ldots z_4$	Motion coordinates corresponding to the masses
c_1/k_1	Spring/damping coefficient with
	i = 1 Parameters of table springs
	i = 2 Parameters of table and baseboard
	i = 3 Parameters of lower mould section and baseboard
	i = 4 Parameters of concrete
	i = 5 Parameters of bellows springs
	i = 6 Parameters of load support
	i = 7 Parameters of knocking bars and baseboard
F_A	Temper head force (compressive force of hydraulic cylinders)
F_F	Mould clamping force (pressure force of bellows springs)
F(t)	Excitation force of vibrator
s_k	Knocking bar spacing

Fig. 1.71:
One-dimensional model of a block machine

models. In order to include the dynamic characteristics of the concrete mix in the vibration models of entire compaction units, for instance, it will suffice, in the first step, to use simple discrete equivalent parameters for the elastic and damping properties. Fig. 1.71 shows the related one-dimensional model of a block machine.

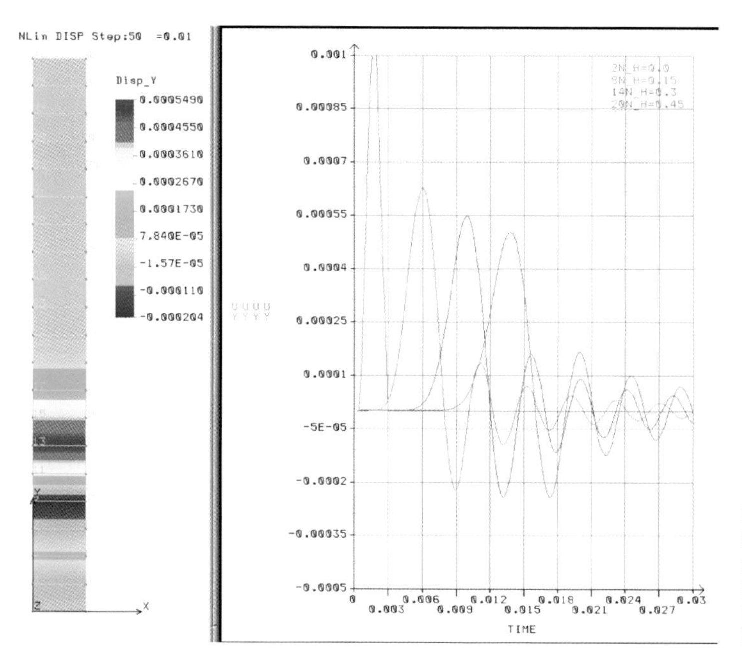

Fig. 1.72:
Pulse propagation in a concrete mix column; simulation using a finite-element model with non-linear time integration

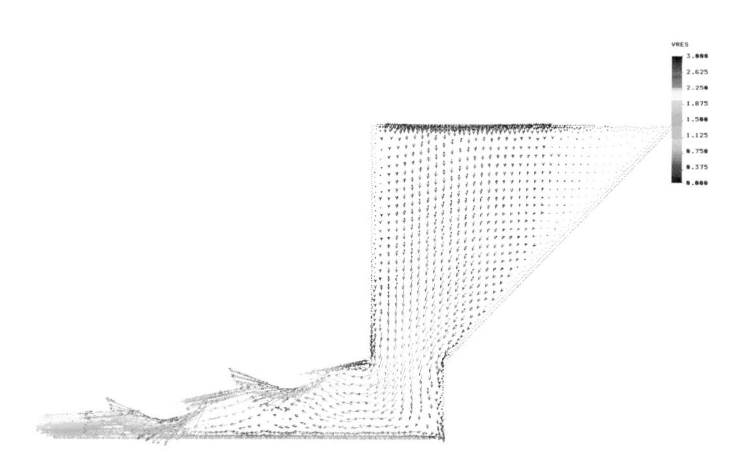

Fig. 1.73:
CFD model of a roof tile machine

In this case, the elastic and damping properties of the fresh concrete are described by a spring constant c_4 and a damping coefficient k_4.

Structural mechanics models are required, e.g. on the basis of the finite-element method (Fig. 1.72), to also consider differences in distribution within the mix, such as the formation of stationary waves in vertically excited concrete columns.

If the processing sequence is a flow process, the application of Computational Fluid Dynamics (CFD) appears useful (Fig. 1.73).

With its corpuscular approach, the particle simulation opens up new options to simulate processes within the concrete mix. Using this method, the concrete mix is represented by a large number of particles connected to each other and to the model walls by contact laws. This makes it possible to observe rearrangement and mixing processes.

The mix properties that can be reflected include the entire range from flow processes of self-compacting concrete to the processing of stiff mixes in block machines [1.117], [1.118]. Reference [1.119] contains a description that concentrates particularly on the simulation of mixing, placement and compaction processes. Fig. 1.74 illustrates the particle simulation model of a concrete placement process in a block machine.

The uniform and quick filling of all mould chambers is crucial to achieve the specified quality standard and to ensure an economical production process. In this regard, the

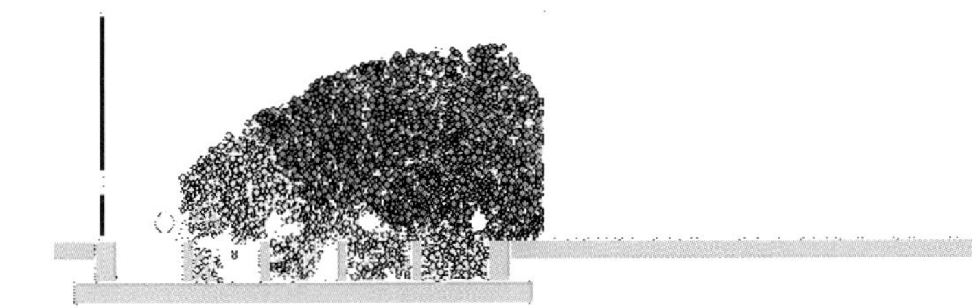

Fig. 1.74:
Particle simulation of the filling process in a block machine

particle simulation helps to recognise movements within the mix that would otherwise be difficult to detect and to analyse the effect of various influential factors [1.119].

1.4.2.2 Dynamic modelling and simulation of production equipment

The dynamic aspect describes the interaction between load and motion in mass systems, and is thus a useful tool for all motion-based systems. For instance, the related findings are applied to the following fields when it comes to designing equipment for the vibratory compaction of concrete mixes:

- generation of vibration
- vibration transfer within the machine
- vibration transfer into the concrete and
- vibration transfer within the concrete

Other areas of application include:

- design of drive systems
- engineering of machine frame systems
- design of machine foundations
- mitigation of unwanted noise and vibration

Two modelling methods are used:

- multi-body systems
- finite-element method

a) Multi-body dynamics (MBD)
In a multi-body dynamics system, rigid bodies are modelled together with the elastic components, joints and forces acting between them. Non-linearities can also be solved in the case of large-scale geometrical modifications. Motion parameters and joint forces, for instance, are displayed in an easy-to-use manner. The system supports the parameterisation of the models. Design studies can be prepared to capture the influence of various parameters. Fig. 1.75 shows the MBD model of a block machine with shock vibration.

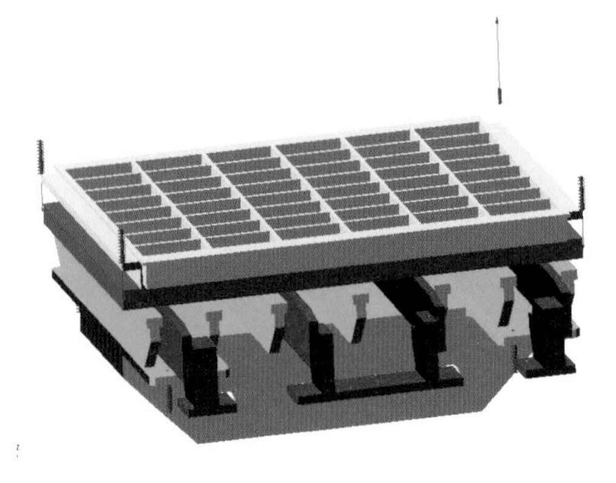

Fig. 1.75:
MBD model of a block machine with shock vibration

b) Finite-element method (FEM)

The finite-element method (FEM) is also suitable for the modelling and simulation of components with complex shapes. This method enables the calculation of natural frequencies, modal components and motion parameters during excitation as well as the analysis of stresses acting during dynamic loading.

Analyses are performed for linear models, such as:

- determination of natural frequencies and modal components
- calculation of stationary vibration by modal decoupling
- modal time integration with impact processes

Non-linear analyses include:

- direct integration of time
- simulations for geometrical non-linearities
- simulations with non-linear material laws

Fig. 1.76 shows the result of a strength simulation performed for a vibrating table of a block machine.

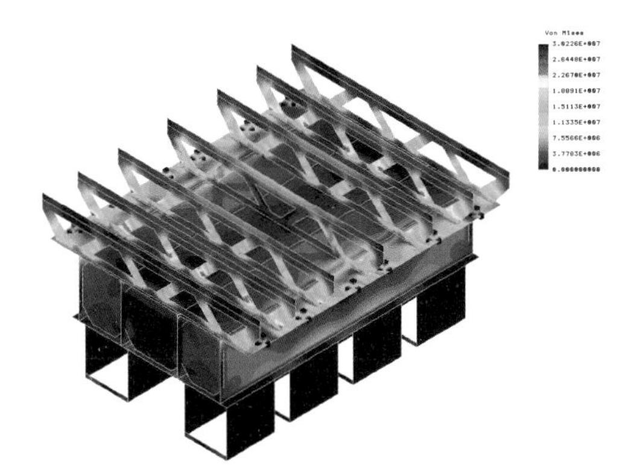

Fig. 1.76:
Result of a strength simulation for a vibrating table

2 Production of the Concrete Mix

2.1 Mixing Facilities

In a mixing facility, raw materials are stored, batched and transported to the mixer, where the actual mixing process takes place, and the finished concrete mix is then discharged. An average-size mixing facility has a production output of 100 m^3/h to 150 m^3/h. Based on the spatial layout of the aggregate bins, there are three basic designs: star-shaped, serial and tower systems.

2.1.1 Star-shaped Systems

Fig. 2.1 shows a schematic representation of a star-shaped system. The individual aggregate grades are stored in open bays with a star-shaped grouping. A centrally located rotary scraper uses a bucket to convey the aggregates to the batching star where they are weighed before being transported to the mixer by a feed hopper or conveyor. The cement is stored in separate bins.

Star-shaped systems are mainly used as mobile mixing units on construction sites. They require a relatively large footprint, and the aggregates are exposed to the weather. These systems are used only to a limited extent in concrete plants owing to the restricted number of possible aggregate grades and because fluctuations in the aggregate moisture content may adversely affect the quality of the concrete.

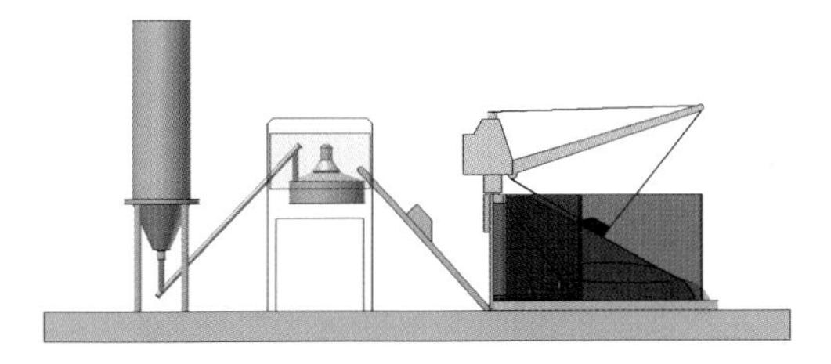

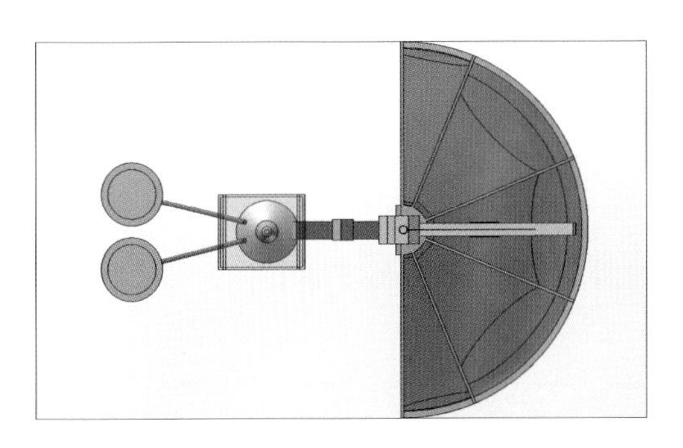

Fig. 2.1:
Schematic representation of a star-shaped system

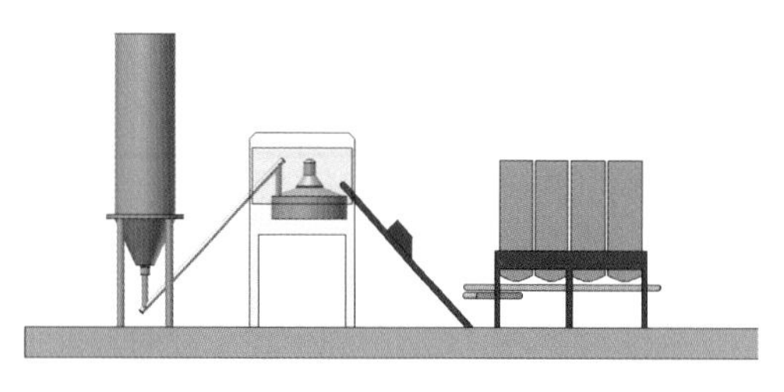

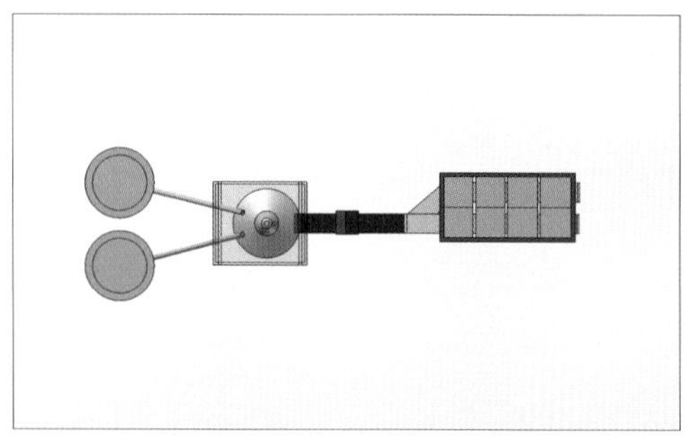

Fig. 2.2:
Schematic representation of a serial system

Fig. 2.3:
Serial system with steel bin and feed hopper

2.1.2 Serial Systems

In serial systems, each aggregate grade is stored in a separate bin, and the individual bins form a row. The material is discharged onto a conveyor system (Fig. 2.2). Again, feed hoppers or inclined conveyors (Fig. 2.3) are used to transport the material to the mixer. The bins may be made of steel or, in stationary plants, also of concrete. In the latter case, they include an underground conveyor system.

Serial systems are frequently used in concrete plants, and the number of possible aggregate grades is larger.

2.1.3 Tower Systems

In tower systems, the aggregates are stored in a silo tower with a number of chambers (Fig. 2.4). The materials are often loaded into the chambers by a bucket conveyor. Discharge and batching to the mixer is gravity-based. Tower systems are highly suitable for concrete plants (Fig. 2.5). The aggregates are weather-protected and can be effectively batched. These systems require a relatively small footprint but a higher initial capital expenditure.

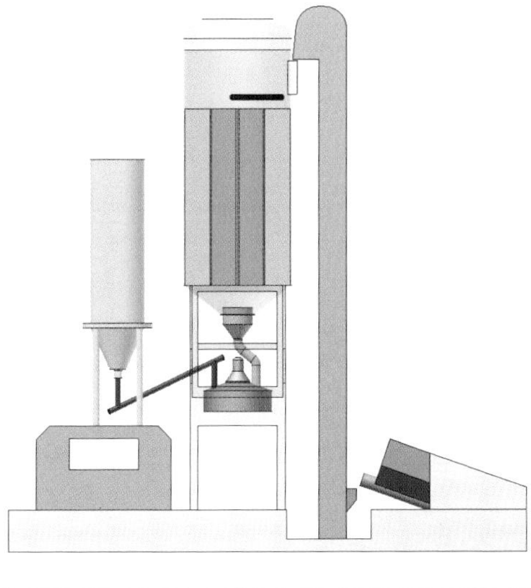

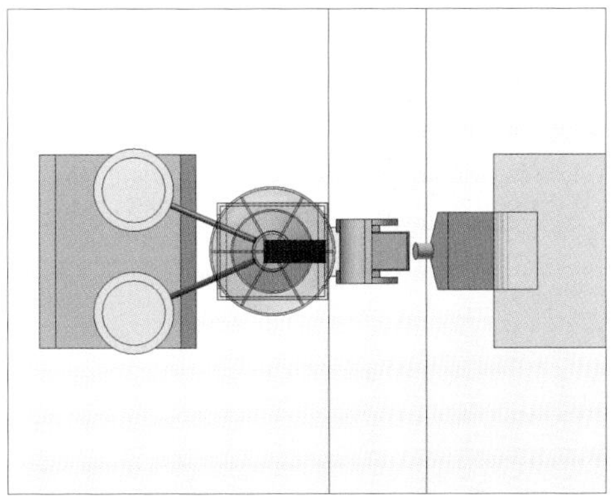

Fig. 2.4:
Schematic representation of a tower system

Fig. 2.5:
Tower system at a precast plant with water treatment and bucket conveyor for concrete transport

2.2 Mixers

Fig. 2.6 shows a systematic overview of the various types of concrete mixers [2.2].

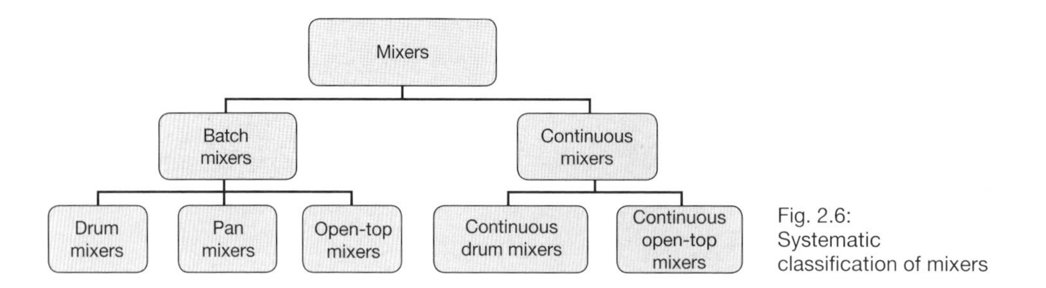

Fig. 2.6:
Systematic classification of mixers

Batch mixers, mainly pan and open-top designs, with more stringent specifications on the mixing quality are used to prepare the mix at concrete plants. Continuous mixers, such as continuous drum or continuous open-top mixers, are not used in this field. Batch mixers used in precast plants mainly include pan mixers and open-top mixers.

Drum mixers (Fig. 2.7) have a rotary mixing chamber that moves about a horizontal or inclined axis. The mix is discharged by tilting or reversing the direction of rotation. The mixing effect is created by the free fall of the materials inside the rotating chamber and is intensified by mixing tools located within the chamber. Drum mixers have a limited range of applications in concrete plants.

Fig. 2.7:
Drum mixer

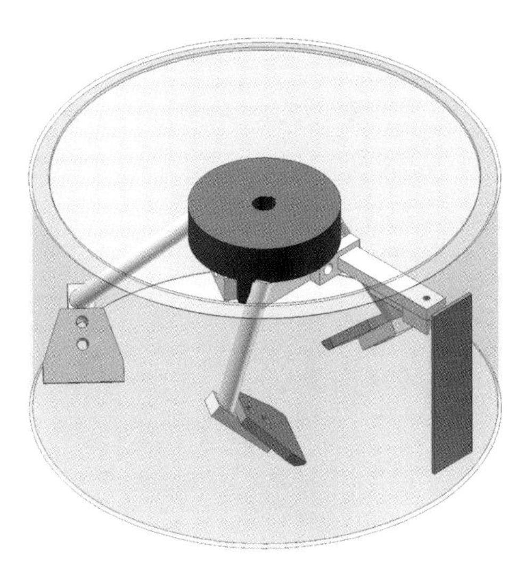

Fig. 2.8:
Basic pan mixer

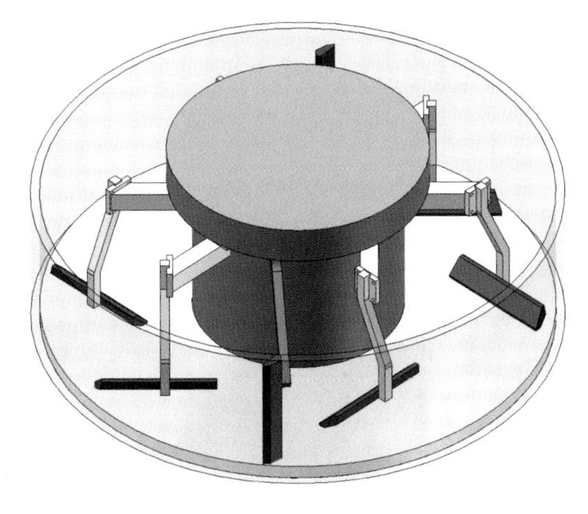

Fig. 2.9:
Ring-pan mixer

Cone mixers are another type of batch mixer. They have a cone-shaped chamber that rotates about a vertical axis and is equipped with rotary mixing screws and paddles. The cone-shaped chamber is well-suited to working with differing filling levels.

2.2.1 Pan Mixers

Pan mixers have a stationary or rotary mixing chamber with a vertical or inclined axis and rotary or stationary mixing tools. Fig. 2.8 shows a basic pan mixer with a stationary mixing pan and concentric rotary mixing tools. The most frequently used types in concrete plants are ring-pan, planetary and countercurrent mixers.

2.2.1.1 Ring-pan mixers

A ring-pan mixer has a circular trough in which several mixing tools rotate (Fig. 2.9). The cylinder axis of the chamber and the axis of rotation of the mixing tools are identical. In this type of mixer, the circumferential speed of peripheral mixing tools is greater than that of mixing tools located nearer the axis of rotation.

Ring-pan mixers have a simple design. The mixing tools are attached to a central drive shaft. In concrete plants, ring-pan mixers are particularly popular for precast element production.

2.2.1.2 Planetary mixers

Planetary mixers are equipped with one or more star-shaped mixing tools. Two rotary movements are superimposed: the stars rotate about their own axis, and the star axes rotate about the chamber axis (Fig. 2.10).

This planetary motion is controlled by mechanical gearing. Due to the constrained motion of the gearing, the stars may engage with each other. The speed of the stars and the planetary speed have a fixed ratio that is determined by the gearing. A second drive option includes separate drives for the stars and for the planetary motion (Fig. 2.11). In this design, the peripheral circles of the stars must not overlap; however, the star and planetary speeds can be selected independently of each other. Planetary mixers ensure thorough mixing and are well-suited to applications in concrete plants.

Fig. 2.10:
Planetary mixer

Fig. 2.11:
Planetary mixer

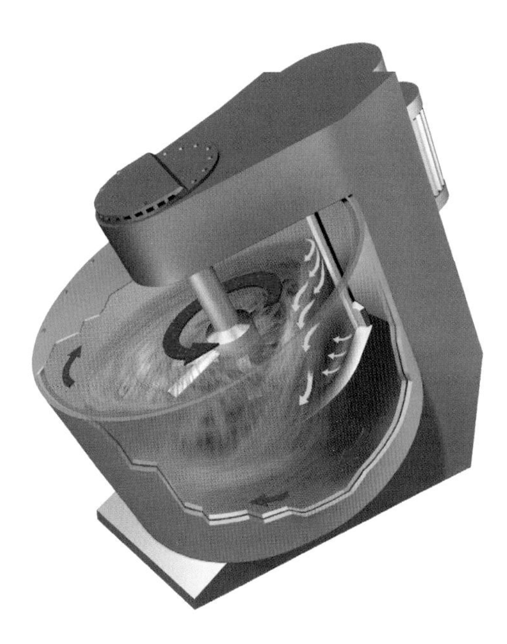

Fig. 2.12:
Countercurrent mixer

2.2.1.3 Countercurrent mixers

The countercurrent mixer (also referred to as Eirich mixer) has a mixing pan that rotates about a vertical or inclined axis (Fig. 2.12). The mixing tools are arranged eccentrically and generally rotate in the opposite direction to the mixing pan. There is a wide range of mixing tool designs. The mixing pan and tools have separate drives with independently selectable speeds. Circumferential mixing tool (whirler arm) velocities of up to 10 m/s are used for mixing concrete.

Countercurrent mixers are primarily used for concretes with fine aggregates, with microsilica or added pigments, and for concretes that must conform to stringent quality standards. These include face concrete for paving blocks, concrete for roofing tiles, SCC and UHPC. Countercurrent mixers are also often used for research purposes.

This type of mixer is more expensive because not only does the mixing tool require a rotary design and drive, but the mixing chamber does as well.

2.2.2 Open-top Mixers

The trough-shaped mixing chamber of an open-top mixer is equipped with mixing tools that rotate about a horizontal axis. Fig. 2.13 shows a single-shaft mixer with individual mixer arms attached to the shaft in a helical arrangement.

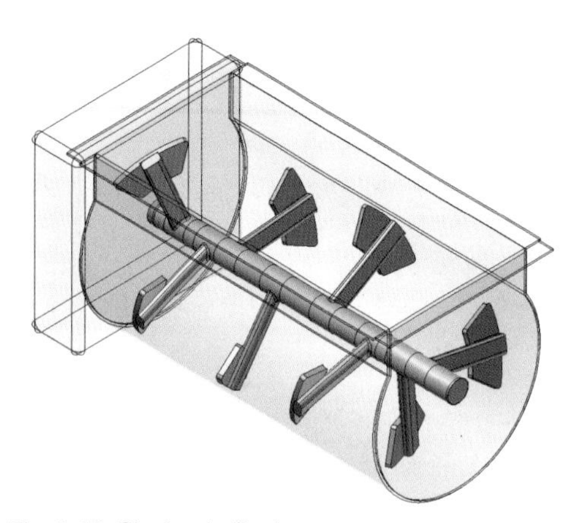

Fig. 2.13: Single-shaft mixer

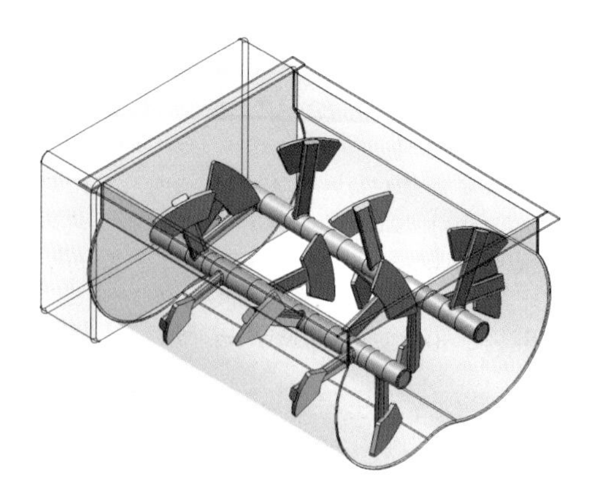

Fig. 2.14: Twin-shaft mixer

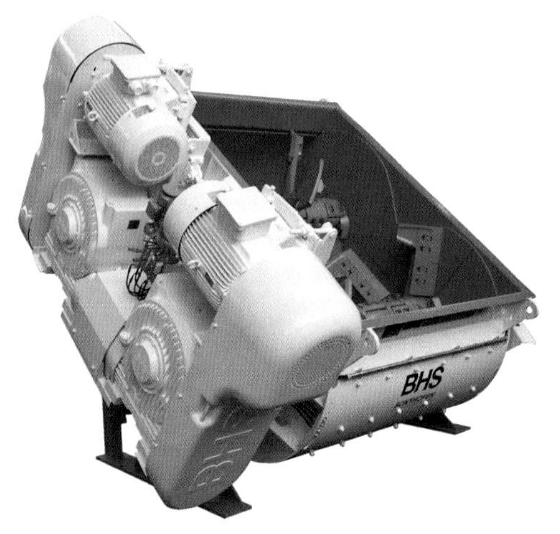

Fig. 2.15: Twin-shaft mixer with drive unit

Twin-shaft mixers are used more frequently in concrete plants than single-shaft mixers (Fig. 2.14 and 2.15).

The two horizontal shafts of a twin-shaft mixer synchronously counterrotate. The mixing volumes of the two shafts overlap in the centre of the trough, thus creating a region with a very high mixing intensity. Twin-shaft mixers are used in ready-mixed concrete plants and precast facilities.

2.3 Quality Control

2.3.1 Assessment of the Mixing Quality

The mixing quality is assessed in accordance with DIN 459-2:1995-11 [2.3] by comparing the fractions of the individual mix constituents in samples taken from the mixed concrete. Special mix formulations are used for this purpose. After a defined mixing time, 20 samples are taken, each with a weight of approx. 15 kg. The constituents of these samples are analysed by washing and screening.

A coefficient of variation is calculated for each mix constituent using the equations below. The coefficient of variation for the fractions of a defined mix constituent (standard deviation of the fractions of the mix constituent relative to the mean value of the fractions of the same mix constituent) is calculated with Equation (2.1):

$$v = \frac{s}{\overline{x}} \cdot 100\ \% \qquad (2.1)$$

The mean value of the fractions of a mix constituent in n samples is calculated with Equation:

$$\overline{x} = \frac{1}{n} \cdot \sum_{i=1}^{n} x_i \qquad (2.2)$$

The fraction of a mix constituent in a fresh concrete sample is calculated with Equation:

$$x_i = \frac{\text{Mass of constituent fraction}}{\text{Mass of fresh concrete sample}} \cdot 100 \qquad (2.3)$$

Thus the standard deviation of the fraction of a mix constituent in n samples is calculated with Equation:

$$s = \pm \sqrt{\frac{1}{n-1} \sum_{i=1}^{n} (x_i - \overline{x})^2} \qquad (2.4)$$

The coefficient of variation characterises the degree of uniformity of the mix. It is a function of the respective mix constituent, the selected filling level of the mixer, the concrete grade, the mixing time and the selected speed.

According to [2.1], the coefficients of variation can be assessed using Table 2.1, for instance for the largest aggregate size.

Table 2.1: Performance classification of concrete mixers according to [2.1]

Performance classes	Coefficient of variation [%]
Standard mixer	< 20
Performance mixer	< 15
High-performance mixer	< 10

Fig. 2.16:
Particle simulation; feed of various aggregate sizes

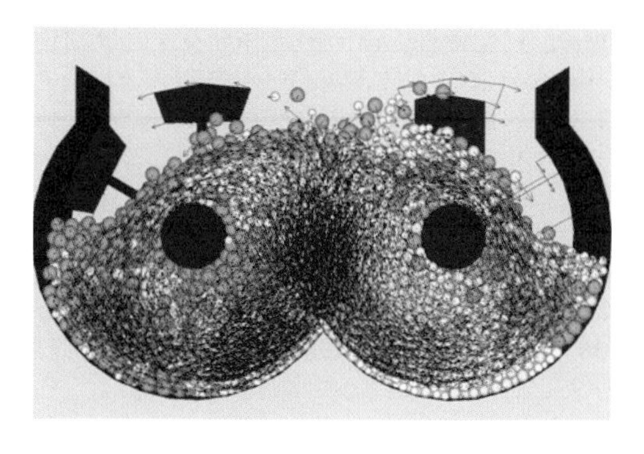

Fig. 2.17:
Particle simulation of a twin-shaft mixer; sectional view with velocity vectors

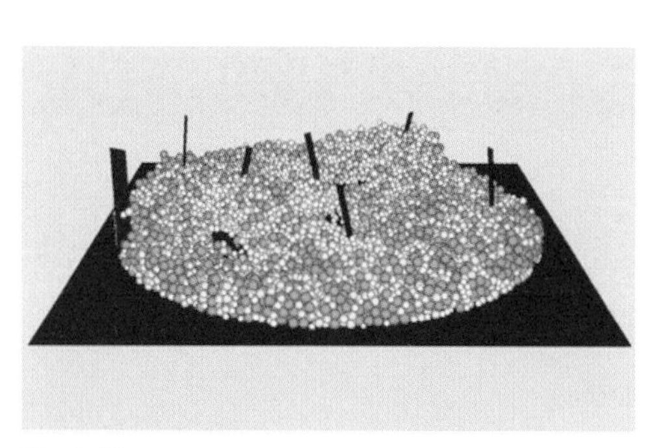

Fig. 2.18:
Particle simulation of a planetary mixer

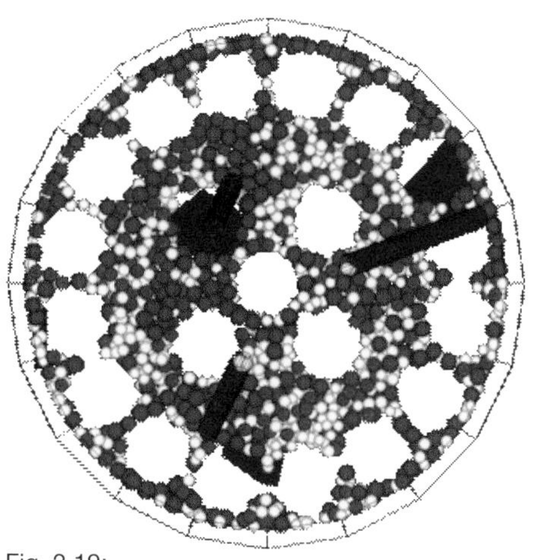

Fig. 2.19:
Virtual sampling of mixed samples

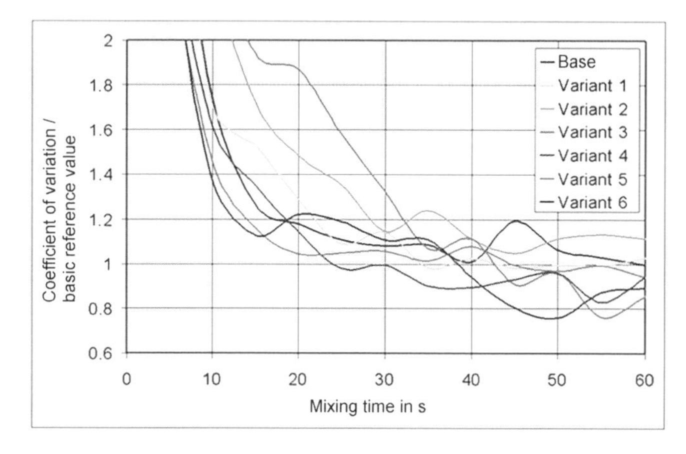

Fig. 2.20:
Coefficient of variation as a function of time for various mixer design options

Particle simulation has recently emerged as a new method of modelling mixing processes. These calculations are based on the discrete-element method. The fractions of the individual aggregate sizes are fed directly into the mixer in the form of particles, and the motion of the material and its degree of mixing are simulated (Fig. 2.16).

Fig. 2.17 and 2.18 illustrate examples of simulated mixer configurations.

One of the advantages of the simulation method is that the coefficients of variation can be determined numerically at any point in time. Virtual samples (Fig. 2.19) are taken out of the mixer, and the coefficients of variation are calculated using the statistical analysis referred to above.

Fig. 2.20 shows the results of this analysis, i.e. the change in the coefficients of variation as a function of time for various tool arrangements and the same mixer. This allows the selection of a design that is best-suited to achieving the intended mixing result. The analysis enables not only assessment of the mixing quality and optimisation of the mixing effect, but also calculation of the forces acting on the mixing tools and the required drive torque. Future versions of this method will also support the design of machine components and improvements in energy efficiency.

ISO 18650-2:2006-04 [2.4] describes a method for assessing the mixing quality that is geared towards the practical aspects of concrete production. Only two samples are taken and analysed for:

- air content
- fines ratio
- particle ratios
- fresh concrete workability
- compressive strength of the hardened concrete

2.3.2 Moisture Measurement

The water/cement ratio is a key concrete parameter that can be adjusted by varying the amount of water added during the mixing process. A change in the moisture content of the aggregates alters the required amount of added water. The moisture content is measured in the mixer to respond to these fluctuations. The moisture sensors, which are often microwave-based, are located in the chamber wall or within special tools (Fig. 2.21).

Moisture measurements can also be carried out during batching of the aggregate. For example, moisture sensors are located near the sand conveyor for roof tile mixes.

Fig. 2.21:
Moisture sensor in a mixer

2.3.3 Mixer Control

The entire mixing system can be operated from a control unit that manages mix designs and process flows, integrates measuring, batching and control procedures and processes data along the entire chain to order handling. The system may also incorporate specific tasks, such as the metered addition of pigments, plasticisers and fibres. To complement the system, an automated mixer cleaning device using high-pressure water nozzles can also be integrated in the control unit.

3 Production of Small Concrete Products

3.1 Overview

The great variety of small concrete products, which continues to grow steadily, enables us to design our living environment so that it not only caters to our needs but is also environmentally friendly and cost-effective [3.1]. Today, small concrete products are used in many areas of construction, a major share being concrete pavers and blocks. German producers offer a particularly wide range of such items (Fig. 3.1).

Depending on their use, small concrete products can be divided into four groups [3.2]:

- Concrete blocks for masonry structures
  These include blocks for walls and ceilings, open-end blocks and chimney facings made of open-structure or structurally impermeable concretes produced with light-weight and/or normal aggregates.
- Concrete products for paving and construction of traffic areas
  These include concrete paving blocks and flags, paving tiles, kerbs and gutters that are generally made of structurally impermeable concretes produced with normal aggregates. In rare cases, such products may also consist of no-fines concrete with normal aggregates.

Fig. 3.1: Concrete blocks in various shapes and sizes [3.2]

- Concrete units for slope stabilisation and/or plot boundary walls
  These include building blocks, miniature palisades and planter boxes that are usually made of structurally impermeable or open-structure concretes with normal aggregates.
- Concrete products for special applications
  These include soundproofing units and foundation blocks made of open-structure or structurally impermeable concretes produced with lightweight and/or normal aggregates.

Cast stone plays a particularly important role in this group of products. This term refers to prefabricated concrete elements whose surfaces are subjected to special requirements in order to achieve a defined design effect (Fig. 3.2). These products include steps, floor tiles, interior and exterior window sills, door and window frames, supports, mouldings, split and embossed facing blocks, wall panels, façade claddings, blocks and seals for walls, garden tables, flower tubs, decorative and honeycomb blocks, and sculptural objects. All these items are covered by DIN V 18500:2006-12 [3.3].

The specifications relating to their performance characteristics are just as diverse as their wide variety, which leads to particular requirements with respect to the design of the associated production processes.

Chapter 4 of this book deals with concrete pipes and manholes, which are also usually grouped under concrete products. The present chapter also describes the manufacture of roofing tiles [3.4] as small concrete products used in building construction (Fig. 3.3).

Fig. 3.2:
Planters

Fig. 3.3:
Roofing tiles with symmetrical "waves" made up of flat and rounded sections of equal width

There are several ways to systematically classify production equipment used to manufacture the wide range of small-scale products. Distinctions are made according to:

- type of concrete product
- special design features of the production equipment
- type of technological production line

The following categories arise with respect to the various types of concrete items they produce:

- block machines
- slab machines
- pipe machines
- manhole machines
- roof tile machines

Specific design features of the production equipment lead to the following grouping:

- board moulding machines
- presses
- vibration moulds

The follow categories arise from the technological point of view (see Section 1.1.2.3):

- Technological lines for stationary production
- stationary production in individual moulds and
- egg laying block machines.
- Technological lines for carousel manufacturing
- carousel manufacturing with individual moulds or pallets,
- carousel manufacturing with board machines and
- carousel manufacturing with pipe-moulding machines.

The following chapter classifies the types of concrete products, taking account of design features and technological characteristics.

3.2 Block Machines

3.2.1 Technological Line

The production of concrete blocks on block machines has become one of the most popular industrial prefabrication methods in the construction sector worldwide. On a global scale, Germany is considered the market leader in the manufacture of equip-

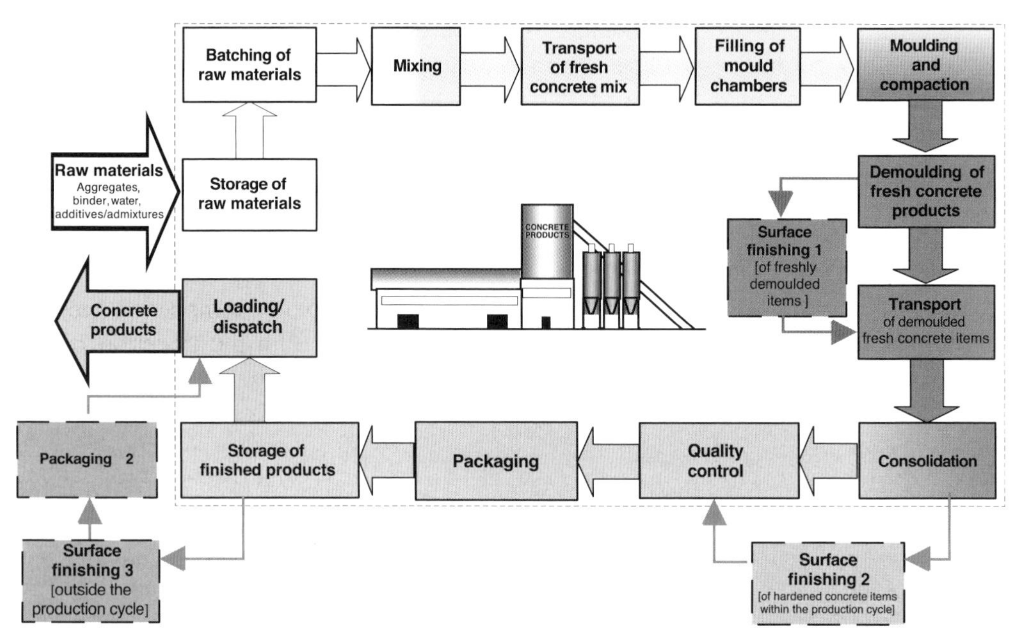

Fig. 3.4: Typical process steps in the production of concrete blocks

ment for concrete block production. Fig. 3.4 shows the individual sub-processes of a technological line used for the manufacture of small concrete products.

In this particular case, stiff concrete mixes are being processed. Table 3.1 lists some examples of mixes for the production of paving blocks.

Table 3.1: Typical mixes for small concrete products:
face mix and core mix for the production of paving blocks

Face mix		Core mix	
Constituent	Proportion [kg/m³]	Constituent	Proportion [kg/m³]
Cement	340	Cement	200
Rock powder	10	Fly ash	100
Sand 0-1 mm	160	Rock powder	15
Sand 0-2 mm	1,090	Sand 0-2 mm	815
Crushed sand 1-3 mm	395	Gravel 2-8 mm	715
Additive (AEA)	1	Gravel 8-16 mm	500
		Additive (CWA)	1.2

Nowadays, fully automated circulation systems are used to implement these sub-processes in a technological line used for industrial production of concrete blocks. Fig. 3.5 illustrates an example of the equipment included in such a circulation system.

At the very heart of the system is the block machine. In the production process, the mould is supported and closed at the bottom by so-called base boards or pallets, which fulfil several functions. During the moulding and compaction phase in the block machine (1), they are initially part of the formwork and thus of the compaction system. Once this stage has been completed, they serve as a base for the transport of the demoulded fresh concrete items along routes (2), (3) and (4) to the curing rack (5). At this point, they become part of the storage system during the curing phase. Thereafter, the base boards serve again as elements of the transport system (4) and (6) en route to quality control (7) and packaging (9). After packaging and transport of the finished products out of the production line (10), the base boards are re-routed to the block machine via a buffer storage facility (8) [3.2]. The base board or pallet is a significant element of the block machine vibration system, which is discussed in one of the following sections. The selection of the pallet material (Fig. 3.6) determines the parameters for spring stiffness and damping of the base board.

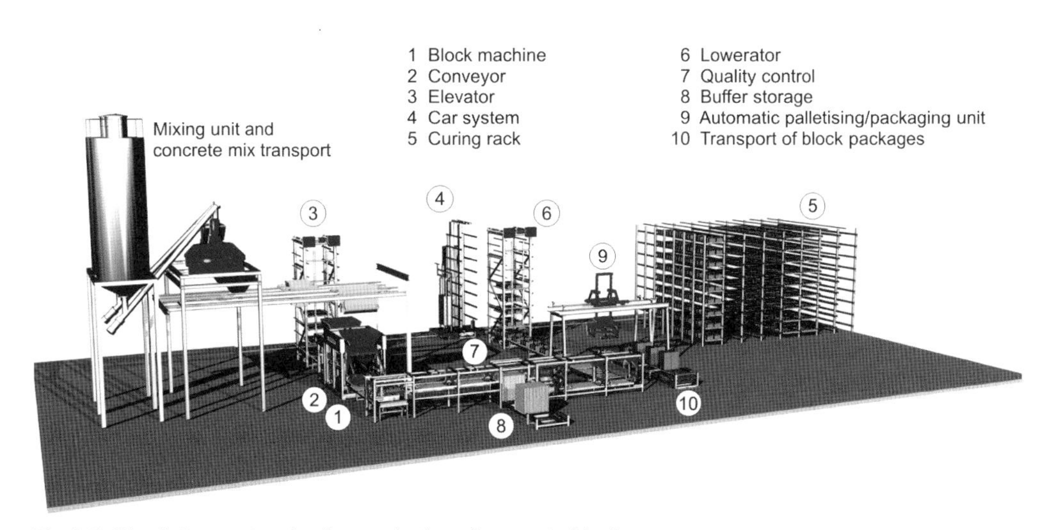

Fig. 3.5: Circulation system for the production of concrete blocks

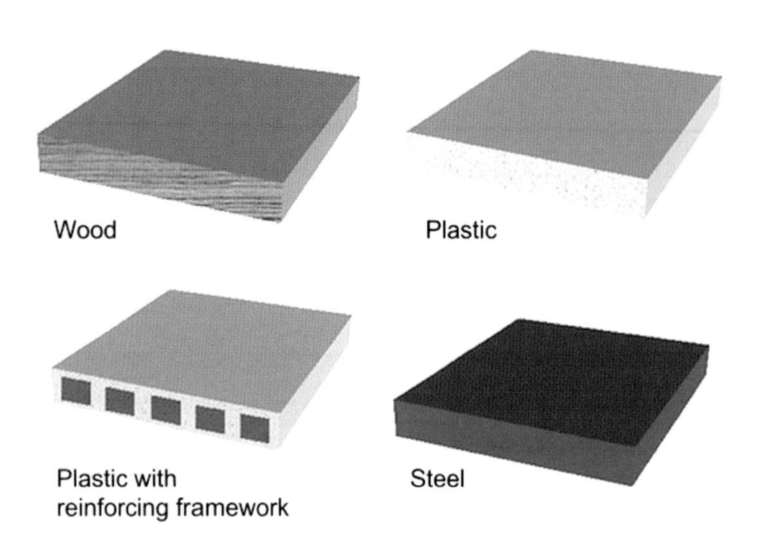

Fig. 3.6:
Base boards made of various materials

Whereas plastic pallets are subject to very minor changes in these parameters over a long production period, hardwood or softwood pallets usually undergo significant changes owing to natural ageing processes and to the prevailing ambient conditions. The multi-purpose function of the pallet requires a number of characteristics to be met by the pallet design.

Moulding and compaction process:	elasticity, damping, mass, dimensions
Transport:	mass, bending strength, dimensions, dimensional stability
Storage:	mass, bending strength, dimensions, water absorption capacity, dimensional stability

3.2.2 Configuration of Block Machines

Block machines are complex, automated pieces of equipment used for moulding and compaction of stiff concrete mixes. The term “Steinformmaschine” (German for “block-moulding machine”) was introduced in the new DIN EN 12629-2 standard [3.5]. In the industry, these automatic systems are mostly termed “board machines”.

Block machines merge three process phases into one compact unit:

- pouring of the concrete mix into the mould
- compaction of the concrete mix
- demoulding of the fresh concrete item

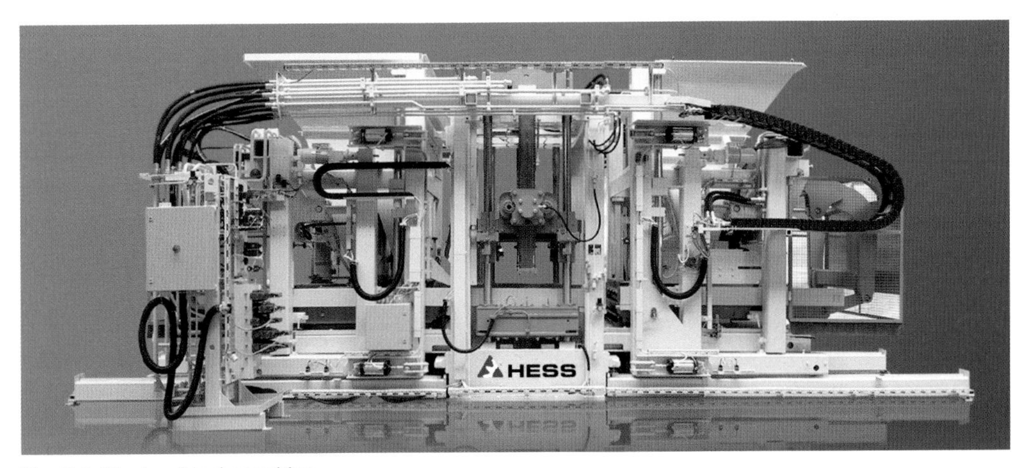

Fig. 3.7: Modern block machine

With regard to their configuration, block machines are complex, dynamic multi-mass systems that respond in many different ways to changes in materials, processes or equipment. Fig 3.7 shows a side view of a modern block machine.

The simplified schematic representation of a block machine in Fig. 3.8 shows a side view that reveals its structural configuration [3.2].

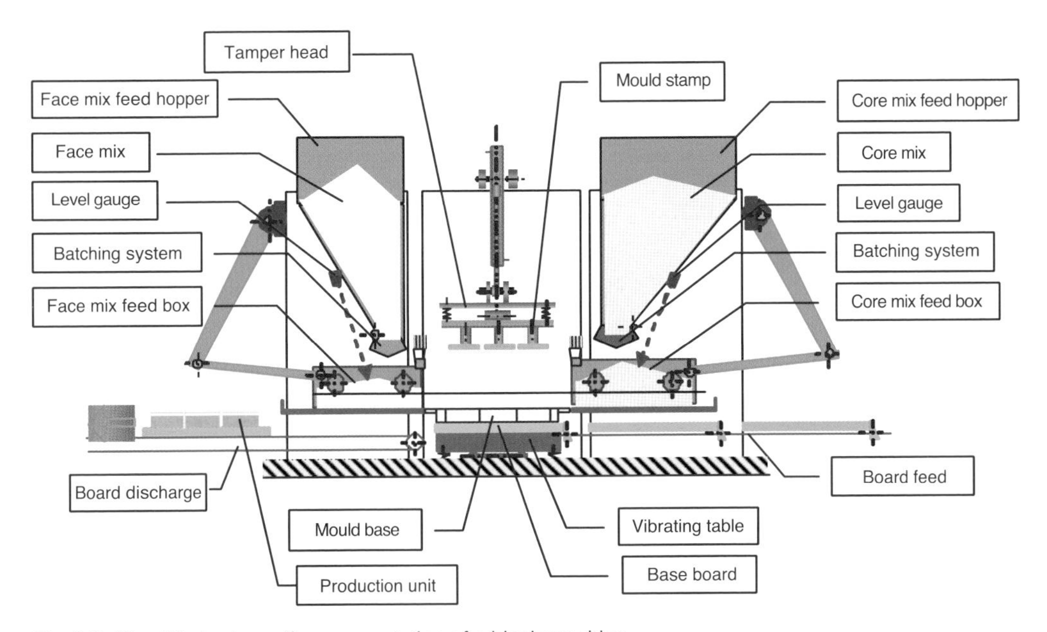

Fig. 3.8: Simplified schematic representation of a block machine

In simple terms, the block machine is composed of

- peripherally arranged components for material feed
- centrally arranged components for moulding and compaction

The interaction between the individual components enables flexible production of a diverse range of concrete blocks.

3.2.2.1 Feed system

One of the basic configurations is a block machine with only a single feed system (Fig. 3.9).

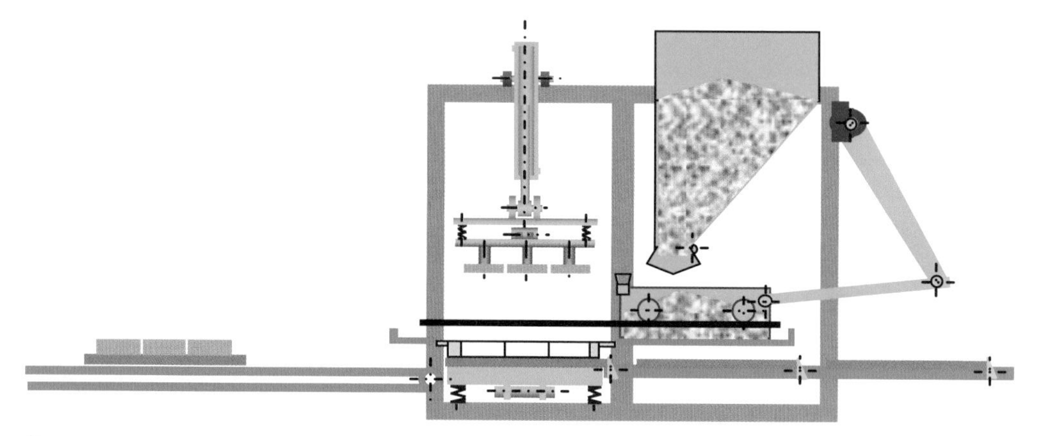

Fig. 3.9: Block machine with core-mix feed unit

This type of machine has a feed hopper and a feed box. Such machines can only perform one feed stage. A typical application is the manufacture of wall blocks.

In order to implement an additional feed stage, block machines are fitted with a second feed system, a so-called face-mix feed unit (Figs. 3.8 and 3.10).

Such a machine can pour two different concrete mixes consecutively into the mould in a single production cycle. A typical application is the manufacture of paving blocks that consist of core and face concretes.

The feed box comprises:

- a box equipped with a vibrating grate
- a wheel-mounted frame

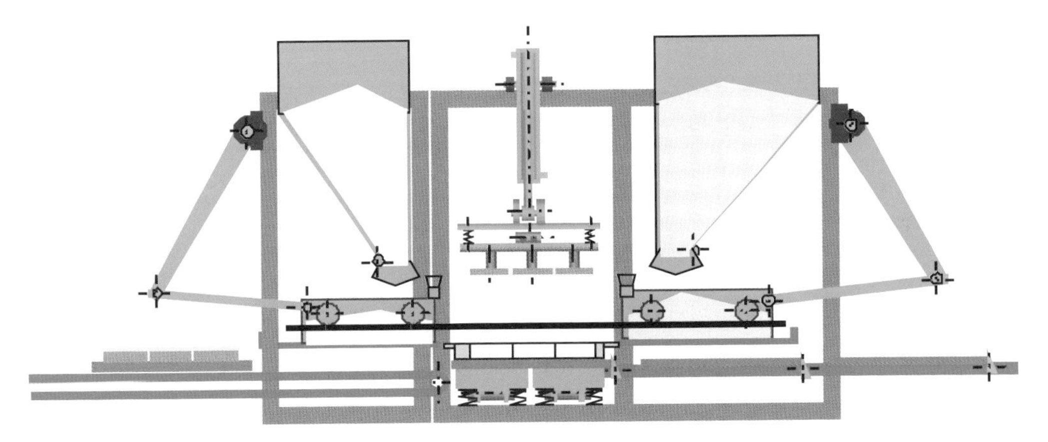

Fig. 3.10: Block machine with core-mix and face-mix feed units and split vibrating table

The bottom ends of the front and rear walls of the feed box are equipped with a scraper. Fig. 3.11 shows a typical feed box design, which is one of a large number of variants available on the market.

References [3.2] and [3.11] analyse commercially available feed systems with regard to their kinematic characteristics and feed behaviour. It is difficult to achieve sufficiently homogeneous filling of the mould or mould chambers with the concrete mix, which is generally stiff. The steadily growing variety of product shapes being manufactured results in many different tasks and processes. In addition, the industry is striving to continuously increase the number of units manufactured per production cycle. The feed sub-process is therefore becoming increasingly important. Reference [3.11] includes a

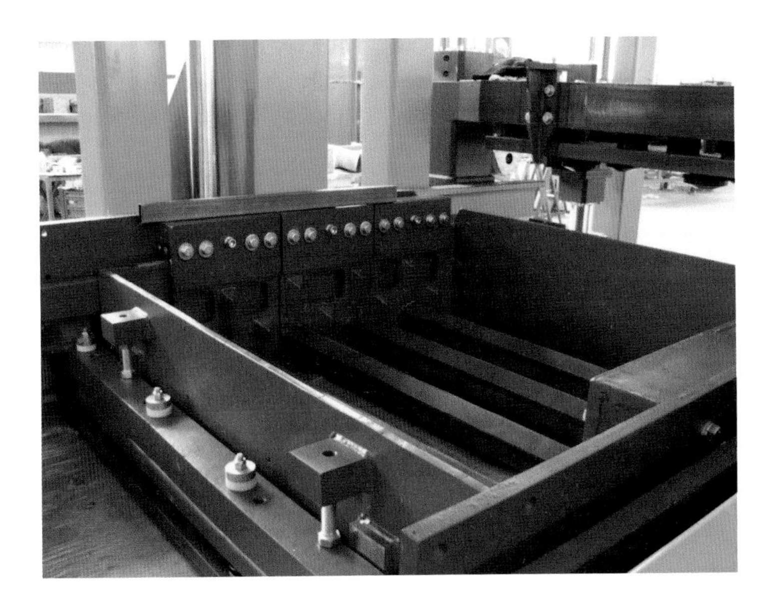

Fig. 3.11:
Core-mix feed box with vibrating grate and scraper [3.2]

Fig. 3.12: Block machine with an innovative feed box

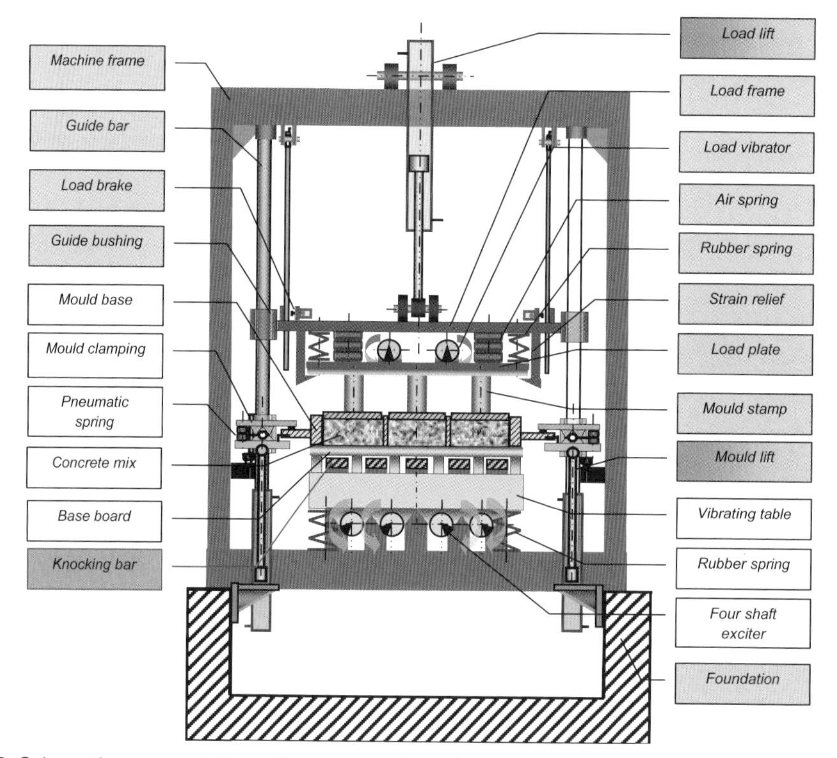

Fig. 3.13: Schematic representation of the compaction unit of a block machine with shock vibration during the main compaction process [3.2]

comprehensive examination of this process based on modelling and simulation of stiff concrete mixes, as referred to in Chapter 1.4.1. This research led to the development of a new technology for filling of concrete block moulds that achieves a more homogeneous product quality and increases productivity (Fig. 3.12).

Reference [3.11] was used in [3.2] to develop methods to quantify and evaluate the achievable filling quality. Approaches to control this sub-process were developed based on an analysis of the feed process as part of the overall technical production process.

3.2.2.2 Compaction unit

The compaction unit of a block machine (Fig. 3.13) comprises the following components:

- load application system and
- vibration system.

a) Mould system

The stamp of the mould is the variable component of the load application system. Its equivalent is the mould base, which is the variable component of the vibration system. As shown in Figs. 3.14 and 3.15, this base comprises a mould insert and a mould frame, which are firmly connected to each other.

The wide variety of intended product shapes necessitates many different designs of the mould base. Such differences in the design also lead to a change in the mass and vibration characteristics of the base, which is also influenced by the potentially asymmetrical cross-section of the mould insert (Fig. 3.16).

Fig. 3.14: Mould base and mould stamp for the production of rectangular blocks

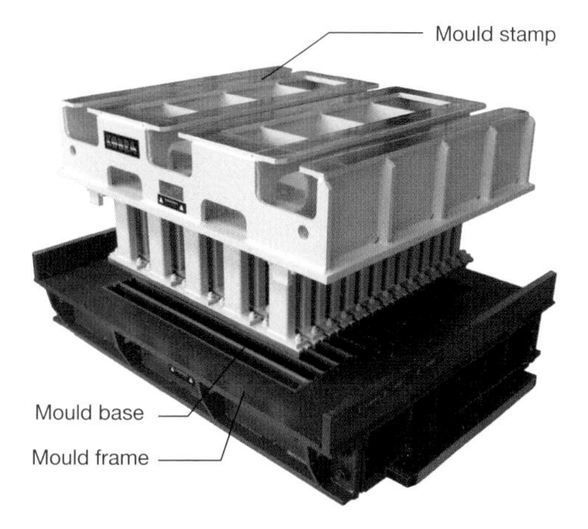

Fig. 3.15: Mould base and mould stamp for the production of concrete edging blocks

Fig. 3.16:
Mould insert with an inhomogeneous, asymmetrical design

During the main compaction process, the vibration table causes the concrete mix fed into the mould to vibrate. At the same time, the mould stamp is pressed onto the mix surface. The compaction in block machines is thus realised by a combination of vibration and pressing. The desired degree of shaping and compaction can be only be obtained if these two actions on the concrete mix are matched to each other.

b) Load application system
Load application systems generally have the same structural design, irrespective of the type of vibration system (Fig. 3.13; blue captions). During the main compaction step, the load is applied to the concrete surface through a gravimetric and/or hydraulic process. Two circular exciters located on the load plate transfer additional vibrations to the mould stamp. Thus, besides the direct impact of the vibration table, vibrations are introduced directly into the concrete surface.

There are many modifications of this basic structural design of the load application system. See [3.6] for further information. The following relationships have been established on the basis of current research and knowledge:

- High dynamic pressure fluctuations at the load/mix interface result in a higher degree of compaction.
- The closer the minimum dynamic pressure fluctuation approaches zero, the higher the degree of compaction (Fig. 3.17).
- A higher load mass leads to an increase in the dynamic pressure fluctuation, and hence to a higher degree of compaction.
- The phase shift between the base board and the load has a strong influence on compaction.

This assumption was also confirmed in [3.6] (Fig. 3.18).

Fig. 3.19 shows the components of the load application system; its contours are highlighted in white. The white arrow indicates the direction of production. The red arrows indicate the piston forces that press the load onto the concrete surface during the compaction process.

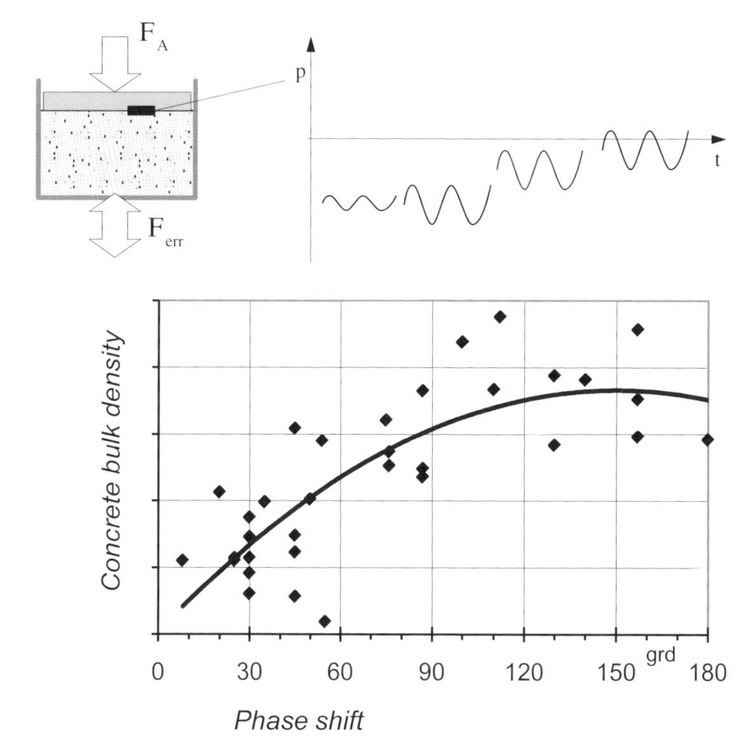

Fig. 3.17:
Diagram showing various states of the applied load pressure

Fig. 3.18:
Influence of phase shift between the vibrating table and the load on the concrete bulk density

Fig. 3.19:
Load application system in the intermediate lowering/ main compaction positions

c) Vibration system
Fig. 3.13 shows the components of the vibratory compaction system in block machines (see yellow captions). The vibration table, which is supported on rubber springs, is excited by an electronically controlled four-shaft exciter system. The table is moved by four rotary unbalance shafts that are driven by servo-motors. Two unbalance shafts rotate synchronously in the respective opposite direction and form a pair, known as a reverse-acting exciter. Fig. 3.20 shows a schematic representation of the operating principle of such a four-shaft circular exciter.

The parallelograms of forces demonstrate that the vertical forces add up, whereas forces acting in opposing directions cancel out each other. Using the electronic control of the mutual angle position of the unbalances, the resulting vertical force component F_V can be set to any value between zero and a defined maximum.

During the initially harmonic motion of the vibration table, impacts are generated both between the vibration table and the base board/mould and between the base board/ mould and the knocking bars. In this regard, it is crucial to set the spacing between the knocking bars and the base board very accurately and uniformly across the entire

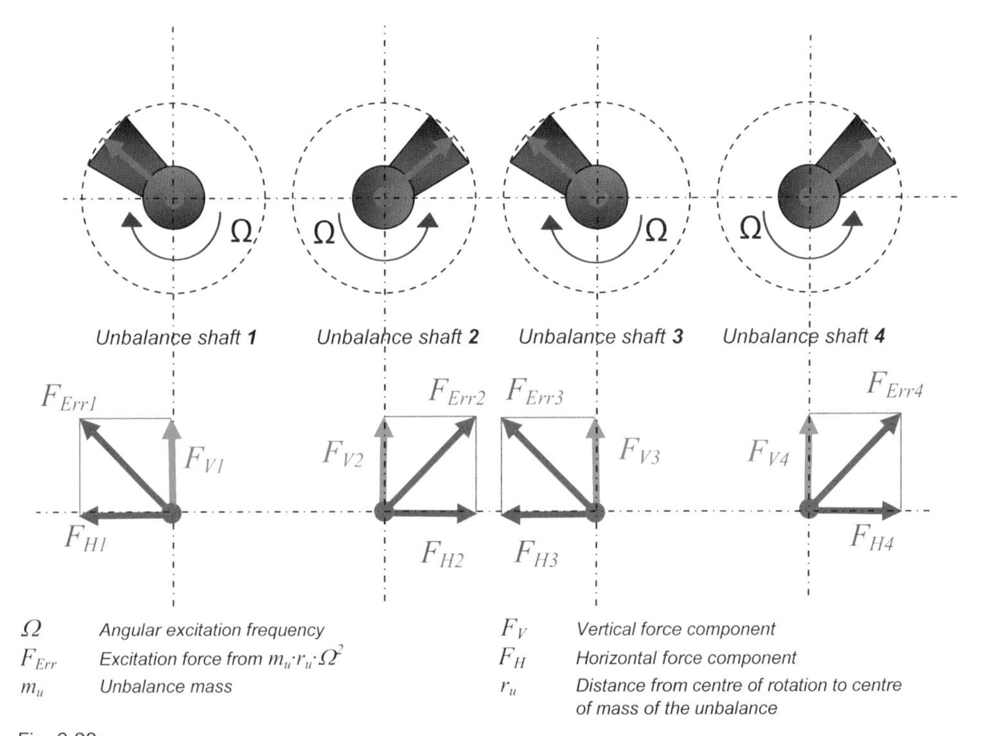

Ω	Angular excitation frequency	F_V	Vertical force component
F_{Err}	Excitation force from $m_u \cdot r_u \cdot \Omega^2$	F_H	Horizontal force component
m_u	Unbalance mass	r_u	Distance from centre of rotation to centre of mass of the unbalance

Fig. 3.20:
Schematic representation of the mechanism of action of the electronically controlled four-shaft exciter

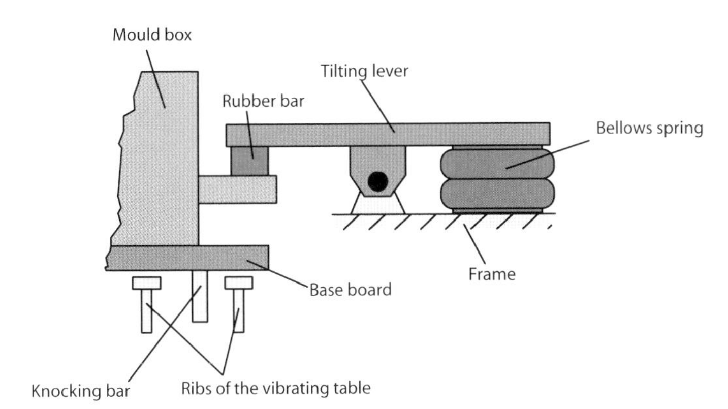

Fig. 3.21:
Pneumatic mould-clamping system using a tilting lever

area. Because pneumatic clamping of the mould (Fig. 3.21) also allows relative motions between the mould and the base board, additional momentums are introduced into the system. In the concrete industry, this is usually referred to as shock vibration.

As investigations referred to in [3.6] have shown, impacts may also occur between the concrete surface and the stamp of the mould during the main compaction phase. They have a significant influence on the final degree of compaction. The components and mechanism of action of shock vibration systems are described in detail in [3.7].

The solution shown in Fig. 3.10 is a special case. It includes a vibration table orientated in a transverse direction relative to the direction of production. This option enables separate control of each of the table halves during concrete feed.

Block machines that use shock vibration are characterised by high-amplitude (in particular pulsed) motion parameters, correspondingly high noise levels and thus a high degree of wear of the compaction system. Fig. 3.22 shows the frequency spectrum measured at the mould of a block machine.

Investigations referred to in [3.8] show that the repeatability of the motion characteristics of a vibration system is not sufficiently accurate in an industrial production environment. A particular issue encountered in this process is the homogeneity of load introduction over the entire base surface of the mould cavity.

One way of preventing these unwanted effects is to generate harmonic movements within the compaction system. In order to avoid relative movements between individual components, the vibration table, base board and mould must be firmly attached to one another to form a single vibrating unit during concrete pouring and compaction. Harmonic excitation of this unit by a specially designed servo-hydraulic cylinder acting as a linear vibrator [3.9] has not become the generally accepted solution in the industry.

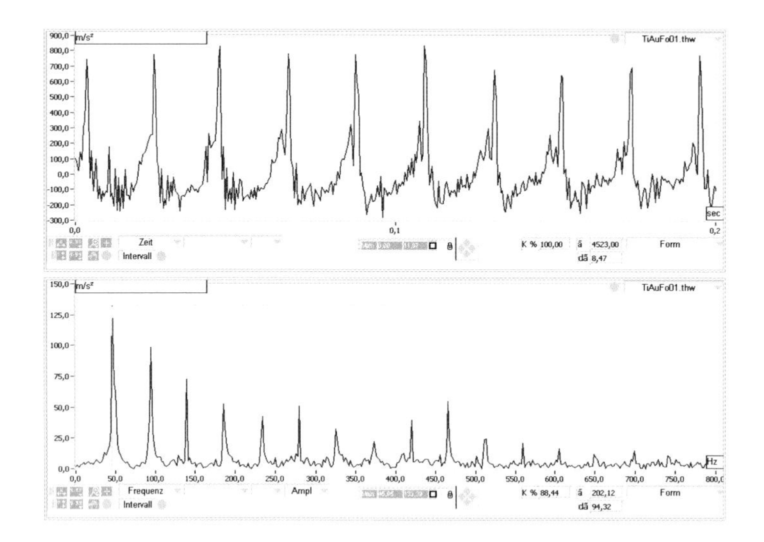

Fig. 3.22: Acceleration/time diagram and acceleration/frequency diagram measured at the mould of a block machine

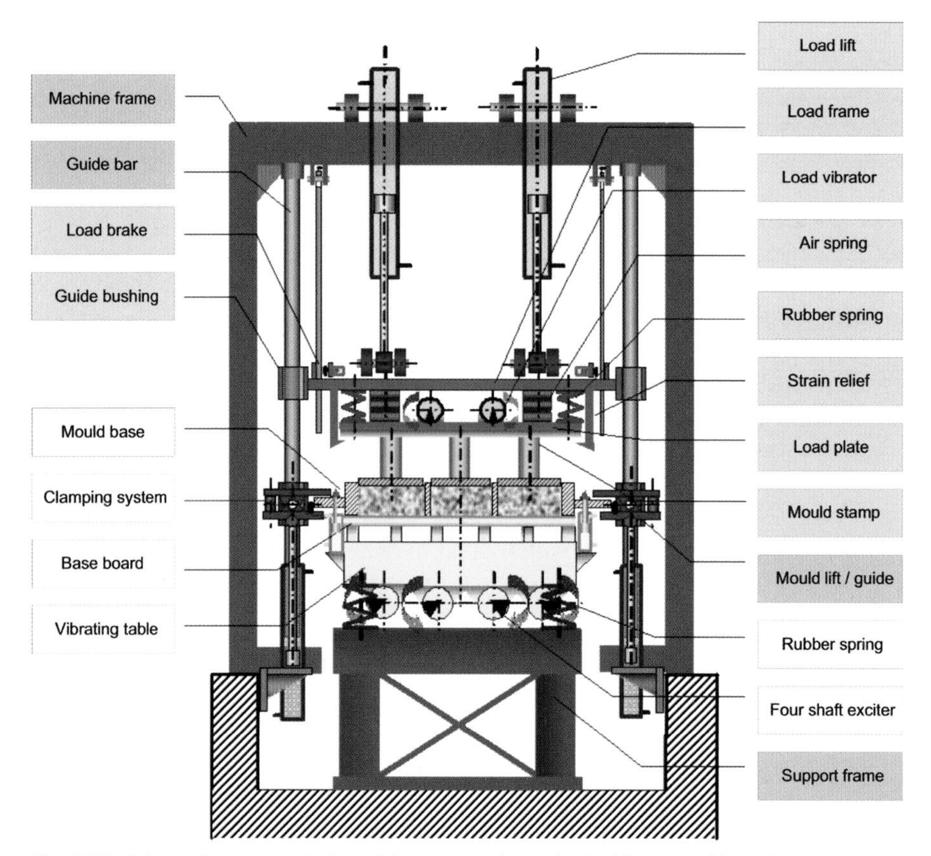

Fig. 3.23: Schematic representation of the compaction unit of a block machine with harmonic vibration during the main compaction process [3.2]

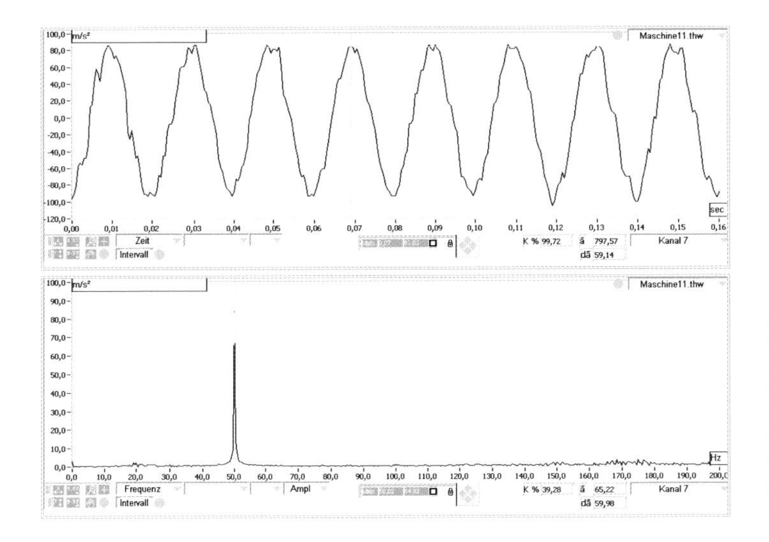

Fig. 3.24: Acceleration/time diagram and acceleration/frequency diagram measured during the first feed vibration on a block machine with harmonic vibration (measuring point positioned vertically on the mould base)

This is due to the significantly higher outlay as well as maintenance and servicing costs for the special hydraulic components.

A new system developed by the Weimar Institute for Precast Technology and Construction uses an electronically controlled four-shaft exciter (Fig. 3.20) to generate harmonic vibration. This system also forms a uniformly vibrating unit that comprises a vibration table, base board and mould base. Fig. 3.23 shows a schematic representation of such a block machine [3.2].

The knocks associated with shock vibration are eliminated to produce an almost harmonic, single-frequency vibration pattern (Fig. 3.24). It should be noted, however, that the excitation forces need to be increased significantly in the harmonic vibration mode in order to achieve a compaction effect comparable to shock vibration. The kinetic moment $m_u r_u$ of the exciter system increases to four times the level measured in conventional systems. As a result, the excitation force may be as high as 800 kN.

This level of compaction energy also enables the production of more massive items that could not previously be manufactured on conventional block machines [3.2]. Harmonic vibration enhances the uniformity of the concrete feed and compaction and thus not only results in a higher product quality, but also significantly reduces noise levels by up to 20 dB. The technical advantages and health and safety benefits of the new system are described in detail in [3.10] and other references.

The vibration table is a steel structure annealed in a stress-free process. It has a special shape (Fig. 3.25) to allow integration of the knocking bars and board feed system.

Depending on the design, compact circular exciters are coupled to the underside of the table or the bearings of the unbalance shafts are incorporated directly in the table structure.

Fig. 3.26 shows the vibration table with a four-shaft exciter and drive separated from the block machine.

Fig. 3.25:
View of the vibrating table and knocking frame of a block machine

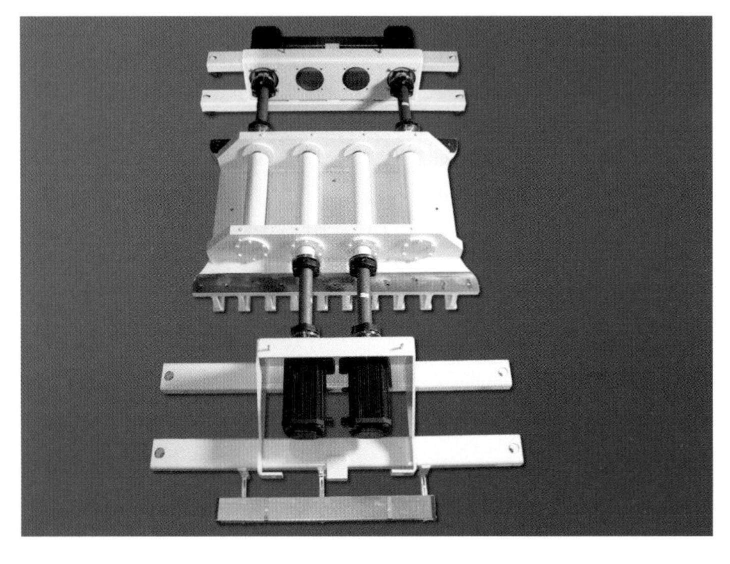

Fig. 3.26:
Vibrating table with four-shaft exciter and drive motors

The vibration table rests on rubber springs supported on the machine frame or a separate support frame.

d) Machine frame and foundation
The function of the block machine shown in Fig. 3.12 is to mould and compact the concrete mix (Fig. 3.4). The method most commonly used for this purpose is vibration. To fulfil the aforementioned function, the vibratory compaction system must be able to oscillate. This means that the points of support on which the vibrating system rests, i.e. the machine frame and the foundation, must be stationary.

The machine frame thus requires a rigid design, which is associated with a certain mass. Section 3.2.3.2 describes how this requirement has been met using the finite-element method (FEM). Dimensioning of the foundation must take account of the rigidity and torsional stiffness of the foundation itself as well as the floor and ground parameters at the site.

3.2.3 Design and Dimensioning of Block Machines

As described in Section 3.2.2, block machines are complex multi-mass systems that respond in many different ways to changes in parameters of materials, processes or equipment. The design and dimensioning of such a machine system must achieve the following:

- implementation of the intended movements and motion parameters as accurately as possible
- dimensioning of the required equipment in such a way that it reliably resists the loads and stresses acting upon it
- mitigate unwanted noise and vibration.

As described in Section 1.4.2, this requires dynamic modelling and simulation of the production equipment, which involves computation of kinetic and kinematic parameters.

3.2.3.1 Motion behaviour

As indicated in Section 1.1.5.2, the motion parameters and excitation frequencies measured at the vibratory compaction equipment are crucial for evaluating moulding and compaction and, consequently, the quality of the concrete products. It is therefore very important to calculate these parameters as accurately as possible beforehand. This can be achieved by applying discrete multi-mass models. Fig. 1.71 shows such a model of a block machine.

One of the motion sequences taking place is shown in Fig. 3.27, which compares calculated and measured movements. In this case, the mould makes contact with the table only during every second unbalance excitation period. A large number of other

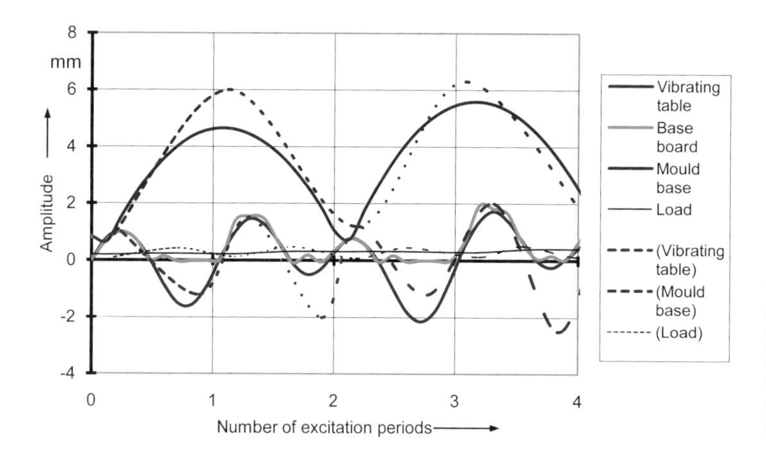

Fig. 3.27:
Motion behaviour of the components of a concrete block machine; comparison of calculated (solid lines) and measured (dotted lines) movements

motion patterns can be distinguished according to their periodicity, time of impact events and engagement of the knocking bars.

The pulsed excitation pattern (i.e. shock vibration) described in Section 3.2.2.2 was found to generate impact-like processes. These trigger inherent oscillation of all system elements capable of vibration, i.e. an entire frequency spectrum (Fig. 3.28).

The authors carried out a large number of vibration measurements of industrial block machines. The results showed that a specific spectrum of acceleration occurs at each of the working masses. Apparently, each individual concrete mix is associated with a specific frequency spectrum that enables optimum compaction of the mix.

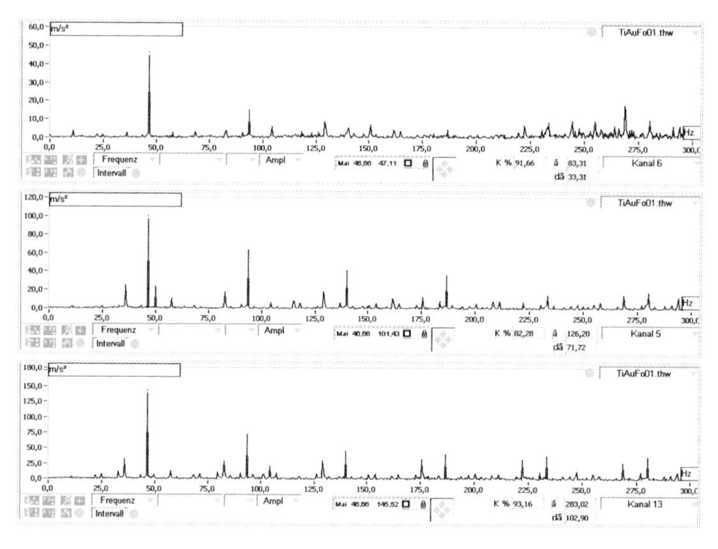

Fig. 3.28:
Frequency spectra of the accelerations measured at the table (bottom), mould (centre) and load application system (top) of a block machine during main vibration

3.2.3.2 Structural design

The structural design of the equipment required for block machines is best carried out on the basis of FEM calculations. As described in Section 1.4.2.2, this simulation method makes it possible to analyse not only natural frequencies, modal components and motion parameters during excitation, but also stresses under dynamic loading. Fig. 3.29 shows such a model of the compaction system of a block machine.

The structural design of the vibration table is of particular importance. It is crucial to ensure the rigidity and torsional stiffness of the table in order to transfer vibration energy into the concrete mix uniformly across the table surface during the moulding and compaction process. This means that the first natural bending or torsional frequency must be at least three to five times the excitation frequency.

The vibration table shown in Fig. 3.13 is supported on the table frame by spring elements that mostly consist of rubber (Fig. 3.30). Such a spring-mounted vibration table can be represented by a vibration model with six degrees of freedom, i.e. six natural frequencies (Fig. 3.31). Of particular interest are the inherent and forced vibrations in the z direction and the tilting vibration about the x and y axes.

The springs on which the table rests ensure its ability to vibrate. The rubber springs used in most cases are distinguished by their static and dynamic characteristics.

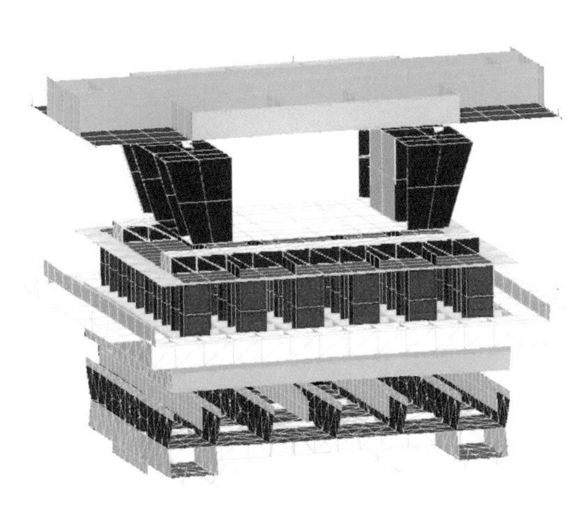

Fig. 3.29: Finite-element model of the compaction system of a block machine

Fig. 3.30: Rubber spring on a vibrating table of a block machine

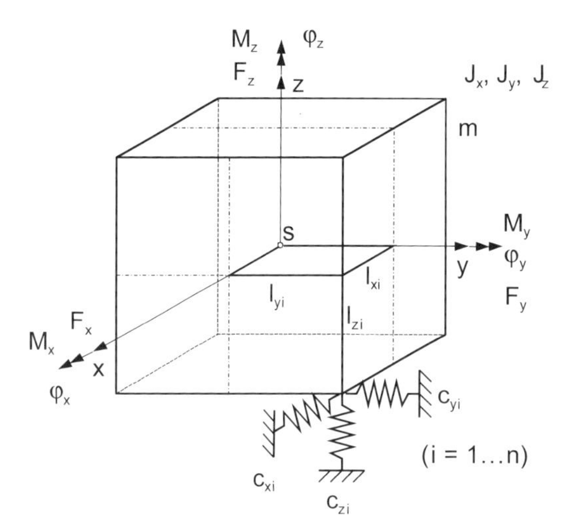

Fig. 3.31:
Vibration model of a rigid machine
- mass m
- mass moments of inertia J_x, J_y, J_z
- spring constants c_{xi}, c_{yi}, c_{zi} in the direc tions of the main axes of inertia
- coordinates l_{xi}, l_{yi}, l_{zi} of their points of application

These primarily depend on:

- frequency
- amplitude of movement
- pre-tensioning
- temperature
- time

At the same time, the springs that support the vibrating table also ensure its isolation against vibration generated by its surroundings.

The selection of the number and type of rubber springs with (as far as possible) identical parameters has a substantial influence on the motion behaviour of the vibration table, and thus on the reproducible quality of the compaction. The basis for the selection of the table springs is described in Section 1.1.4.3. Supercritical operation of the

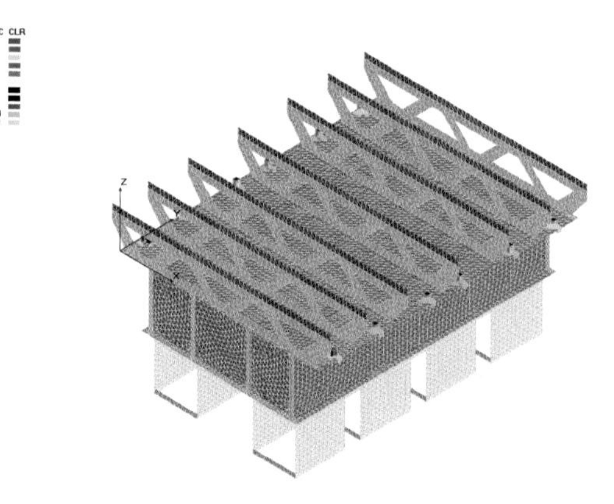

Fig. 3.32:
Finite-element model of the table with springs

system is assumed, i.e. a calibration ratio of $\eta = 3$ to 5. On this basis, the corresponding system parameters are calculated for its movement in the z direction.

The following paragraphs explain an example FE analysis of a vibration table using the model shown in Fig. 3.32, which consists of:

- extensive shell elements representing the sheet metal components
- bar-shaped elements representing the springs.

The first ten natural frequencies were calculated (see Table 3.2) for the model shown in Fig. 3.32.

Modal components 1 to 6 represent the rigid-body modal components of the vibration table. The first natural torsional frequency occurs at 338 Hz (Fig. 3.33). Since this frequency is considerably higher than the excitation frequency of 60 Hz, the vibration table is rigid relative to the latter. Under these conditions, the natural frequencies of the rigid-body movements of the table can be compared with the excitation frequency, which means that the calibration ratio η can be controlled. In this example, the condition of $\eta = 3$ to 5 is fulfilled except for the natural tilt frequency about the y axis and the natural vertical frequency in the z direction.

In the next step, the forced vibration parameters of the vibration table can be calculated depending on the excitation frequency. For this purpose, the excitation forces are introduced into the bearing points of the unbalance components (Fig. 3.32). Due to the assumed unbalance excitation, these forces are frequency-dependent.

Table 3.2: Natural frequencies of the vibrating table

Natural frequency no.	Frequency [Hz]	Comments
1	9.3	natural horizontal frequency in the y direction
2	9.5	natural horizontal frequency in the x direction
3	11.7	natural tilt frequency about the z axis
4	17.0	natural tilt frequency about the x axis
5	22.0	natural vertical frequency in the z direction
6	31.1	natural tilt frequency about the y axis
7	338.0	first natural torsional frequency
8	338.4	first natural bending frequency
9	358.2	second natural bending frequency
10	361.7	bending of the table bars

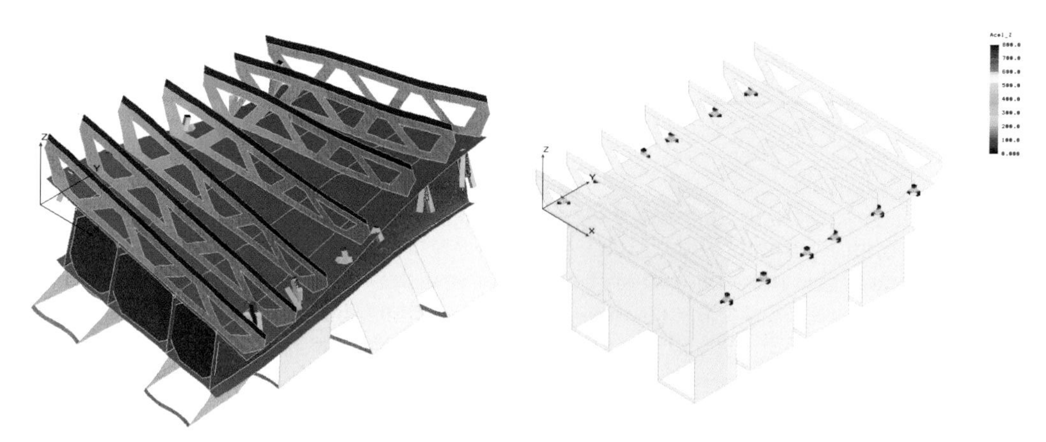

Fig. 3.33:
First torsional modal component at 338 Hz

Fig. 3.34:
Vertical acceleration of the vibration table at 60 Hz

Fig. 3.34 shows the vertical acceleration values for the vibration table at 60 Hz.

On this basis, the design and dimensioning of the vibration table, i.e. of the excitation forces and thus the vibration exciter as such, can be carried out whilst taking into account the vibratory compaction parameters referred to in Section 1.1.5.2. This step determines the mechanical loads acting on the table structure, which form the basis for the verification of the strength parameters of the table, i.e. the calculation of the existing stresses. On the basis of the material selected, the vibration table is then designed and dimensioned from a structural point of view. Fig. 3.35 shows the distribution of stresses on the vibration table at an excitation frequency of 70 Hz and an excitation force of 400 kN.

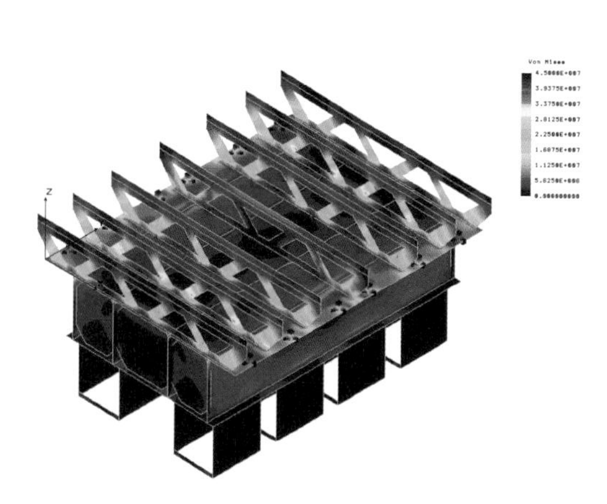

Fig. 3.35:
Existing von Mises stresses for an excitation frequency of 70 Hz and an excitation force of 400 kN

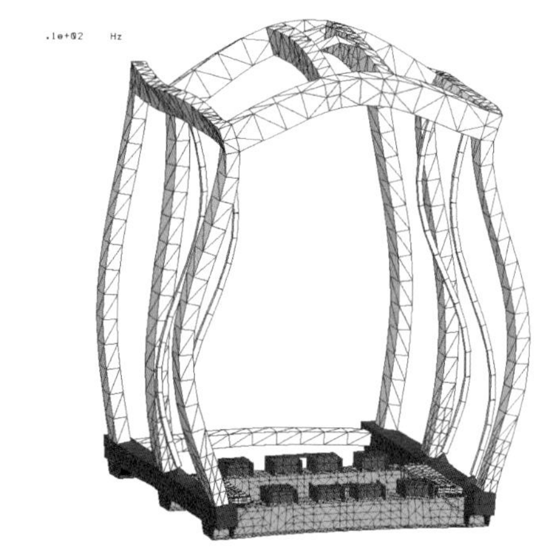

Fig. 3.36:
Calculation of a modal component of the deformation of block machine frame structures using the finite-element method

As demonstrated above for the vibration system, the same approach using the aforementioned calculation methods can also be applied to the structural design of all parts and components of block machines, such as the frame structure (Fig. 3.36) and the foundation.

3.2.3.3 Foundations

The purpose of the foundation of a block machine is to avoid propagation of the mechanical vibration created by the operating machine to its surroundings in order to prevent undesirable hazards and adverse effects on employees, plant and machinery, and buildings. There are several foundation designs that fulfil this isolation requirement in different ways [3.12]. Typical foundations for block machines are listed in Table 3.3.

Table 3.3: Common foundation designs for block machines

	Ground foundation	Isolated foundation pad
Schematic drawing		
Description	machine frame on massive foundation pad, lateral isolation, foundation base on the ground	machine frame on foundation pad, foundation pad on spring elements or elastic layers in an isolated trough
Natural vertical frequency	25 – 40 Hz	5 – 15 Hz
Vibrational stress on machine and frame	medium	medium
Transfer of vibration to surroundings	higher, depending on the design of the foundation	low
Permeability for table excitation frequency $V_d = \frac{\hat{F}_{ground}}{\hat{F}_{exc}} = \frac{1}{1 - \eta^2}$	0.3 – 2.0	0.01 – 0.1
Costs	high	very high

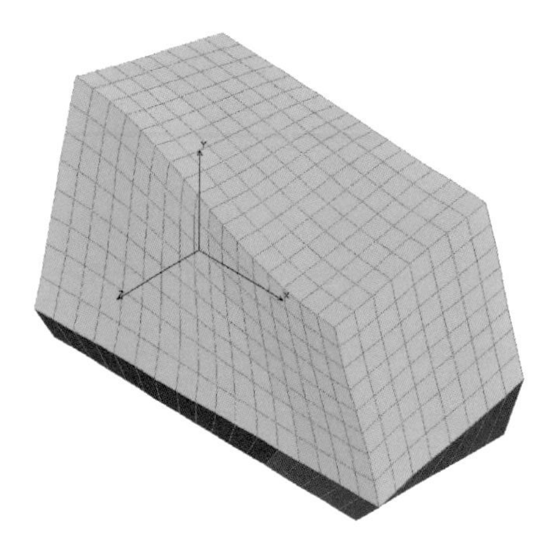

Fig. 3.37: Modal component of the deformation of a solid concrete block at 210 Hz

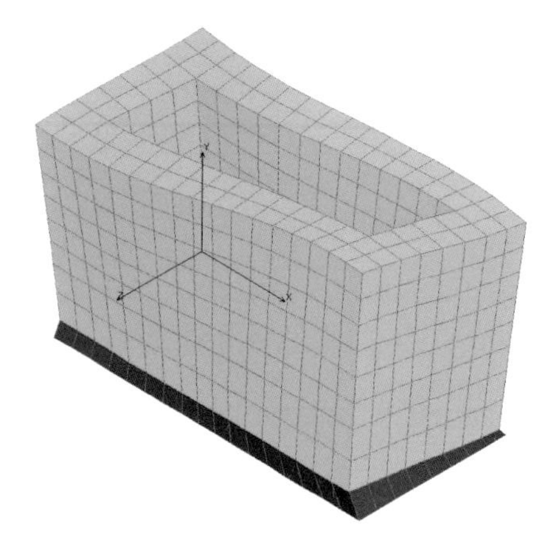

Fig. 3.38: Modal component of the deformation of a foundation with pit at 210 Hz

The foundation must be designed so that it is rigid at the excitation frequency. Its rigidity is assumed if the first natural bending frequency amounts to at least three times the highest excitation frequency. At excitation frequencies between 30 and 100 Hz, the first natural bending frequency of the foundation must be greater than 300 Hz. The foundation should be calibrated to low frequencies, which means that the natural rigid-body frequencies are significantly below the lowest excitation frequency.

The simplest and most favourable foundation design is a compact solid block with a low length-to-height ratio. Fig. 3.37 shows the first modal component of the deformation of such a solid block. For maintenance reasons, the foundation often includes a pit. Fig. 3.38 shows the deformation characteristics of a foundation with a pit.

3.2.4 Quality Control

3.2.4.1 Aim and purpose of quality control measures

This section first outlines the principles of quality control, which are then applied to the individual concrete products.

Assuming that the quality criteria for the raw materials described in Chapter 1.2 are met, quality control is the part of quality management that deals with monitoring of production steps critical to quality and checking of the finished products (see also Chapter 1.3). All quality control systems aim to ensure that products are manufactured to a uniformly high quality standard as effectively as possible.

Quality specifications for concrete products and precast elements result from their shape, intended surface finish (colour and texture) and performance characteristics, such as dimensional accuracy, strength and durability.

The most important quality parameters and associated testing methods are covered by national standards and internal guidelines pertaining to the individual product groups. Depending on the specifications defined by the customer, this may lead to a situation where internal quality guidelines impose requirements that are stricter than those stipulated in the applicable national standard (e.g. height tolerances for concrete pavers).

3.2.4.2 Basic principles of quality control

Quality control is subject to two basic principles:

1. Control of the finished, cured products
2. In-process control of the raw materials, intermediate products and production steps relevant to quality

The benefits and disadvantages of these principles are listed in Table 3.4.

The first quality control principle mentioned above has become standard practice. End-to-end in-process control is becoming increasingly important for selected quality parameters and has already been implemented for some sub-processes.

Table 3.4: Finished-product quality control versus in-process quality monitoring

	Advantages	Disadvantages
Control of finished, cured products	Quality parameters with testing methods clearly specified in standards	Defects are identified only for the finished product: • root cause of defects is often difficult to identify • testing methods are generally time-consuming and costly, destruction of test specimens • size of sample to be tested relatively small
Monitoring of production steps critical to quality	Defects are identified during or immediately after the relevant production step: • root cause of defect easier to identify • fast feedback into the process to rectify causes of the defect • reduction of the reject rate and easier re-processing requiring a lower amount of energy • under certain circumstances, possibility of checking all batches produced	In-process quality control systems for the manufacture of concrete products and precast elements are still in their infancy: • existing quality control standards applicable only if certain conditions are met • comprehensive research and development effort still needed in some areas • high cost, application of the systems may not be welcomed by employees

However, there is still a significant need for research and development in this field.

In many cases, in-process quality control does not monitor the quality parameters specified in the standard but parameters from which a product quality statement can be derived either directly or indirectly.

3.2.4.3 Possible solutions and selected examples of in-process quality control

Important parameters relevant to in-process quality control are:

- dimensional accuracy, shape
- concrete strength, strength indicators
- surface quality (colour, texture, roughness)
- workability, consistency of the concrete mix

The following requirements apply to any in-process quality control procedure:

Table 3.5: Methods for testing relevant quality control parameters

QC parameter	Method/solution	Comments
Dimensional accuracy, shape	non-contact measurement of the shape, e.g. using laser or ultrasound systems and image processing	first systems in use under real-life conditions, need for further development regarding the adjustment to various transport/handling systems and other product groups
Concrete strength	indirect determination of required QC parameters via fresh concrete bulk density indirect determination of fresh concrete bulk density via weighing and non-contact height measurement (laser) direct determination of strength using ultrasound	(as above) significant need for research and development
Surface (colour, texture, roughness)	check of texture and colour using optical methods (optical system coupled with CCD line-scan camera, novel colour sensors) determination of roughness, e.g. using laser-based methods	high cost, viable only for high production outputs and/or premium-finish products highly susceptible to faults under changing optical conditions; significant further development needed
Workability of the concrete mix	check of workability under the impact of dynamic compaction parameters	very comprehensive research required

- The test should be non-destructive.
- The production process should not be disturbed, or only to a minimal extent.
- The test should yield a result as quickly as possible so that immediate interventions in the production process are possible in the case of quality deviations using existing setting options or control algorithms.

In accordance with the requirements specified above, non-destructive and – if possible – non-contact testing methods (Table 3.5) are appropriate for most in-process control environments. In addition, as cycle times or rates during production are often very short (concrete products: < 10 s, roofing tiles: > 100 pcs./min), a large amount of data needs to be processed.

3.2.4.4 Integration of state-of-the-art process control systems in quality control

Within the information processing structure of a concrete plant, state-of-the-art process control systems form a certain type of nodal interface between the operator, the collection of operational data and the various components for process control and monitoring. In other words, the process control system closes the gap between the superior PPS system (factory data administration) and the level of the individual process control units. It is indispensable for the implementation of an end-to-end information processing structure because it creates the IT infrastructure for:

- novel options for data collection
- development of correlations between product quality, material and equipment parameters
- optional transfer of feedback into the process (which is a requirement to create quality control loops)
- centralised monitoring of production at distributed production sites

3.2.4.5 Quality criteria

A crucial factor that influences the quality of concrete products is moulding and compaction of the concrete mix. As described in Section 1.1.4.2, vibration is the most commonly used method for this purpose. Apart from material-related aspects, there are a number of other parameters that determine product quality. Section 1.1.5.2 provides a description of these parameters, grouping them into several classes.

Unfortunately, it is not yet possible to measure internal compaction parameters during the moulding and compaction process. For instance, attempts to determine the change in the bulk density of fresh concrete as a termination criterion during the compaction process have been unsuccessful so far. What can be measured, on the other hand, are the actions on the edges of the concrete mix in the form of motion parameters, using appropriate acceleration sensors.

For this reason, the approach to quality control currently feasible in the manufacture of concrete products is to determine those influential parameters, in laboratory and pilot-scale tests, that ensure the intended moulding and compaction, to compare them with existing parameters by carrying out measurements at production lines under real-life conditions, and to derive suitable instrumentation and control solutions for the equipment system.

In block machines, shock vibration is an additional factor for ensuring the required effectiveness of compaction. A comprehensive description of the related details is given in Section 3.2.2.2. As demonstrated in Section 3.2.3.1, shock vibration is associated with typical spectra. For each concrete mix and product shape, specific frequency spectra appear to exist that enable the optimum moulding and compaction of the concrete mix.

Summary
As regards vibration technology, quality control measures can be grouped in three stages:

1st stage:	One-time measures to design and adjust the compaction equipment, supported by vibration measurements, laboratory tests and models.
2nd stage:	Monitoring of the vibration impact (display of spectral pattern) and, if applicable, extension to the monitoring of input and output parameters.
3rd stage:	Operation and control of the vibration system on the basis of vibration and/or product data.

3.2.4.6 In-process quality control measures
The in-process quality control system referred to in Section 3.2.4.2 is becoming increasingly popular. Some examples are listed below.

a) Measurement of block height
Measurement of the block height (Fig. 3.39) is a quality control system to determine the height of concrete products immediately after demoulding, using a non-contact method. Its advantages are:

- end-to-end production monitoring
- no interruption of the production sequence
- quick and early response to any occurring quality deviations
- prevention of production waste that is difficult to dispose of – defective products can be separated early on (at the green concrete stage) and fed back to the mixing unit
- more effective detection of defects using archived quality data.

Each full pallet passes the height measurement system (3) separately. The block height is measured in selected areas of the pallet. Another sensor captures deviations of the pallet thickness from the defined target value. The height is calculated immediately for selected rows of blocks and is then displayed on a monitor in a designated colour code.

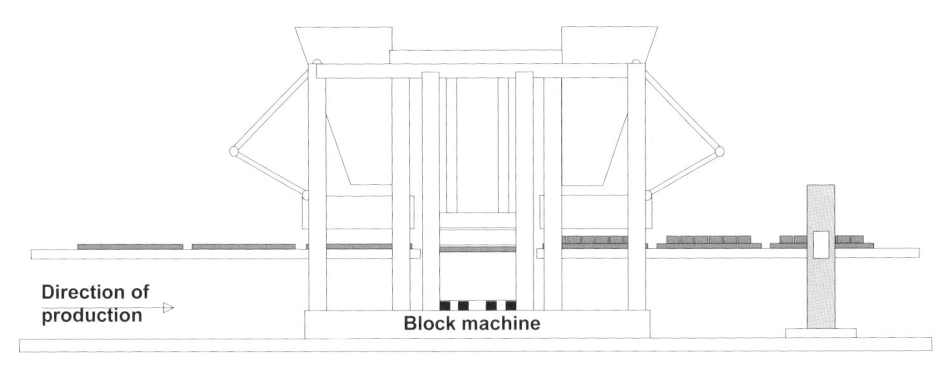

Fig. 3.39: Principle of block height measurement

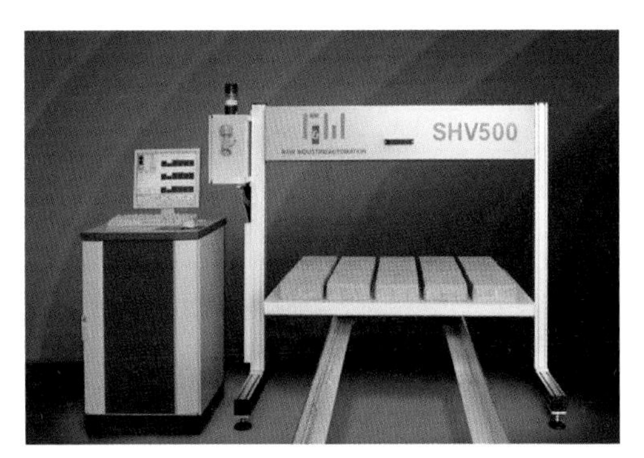

Fig. 3.40: Laser systems for height measurement

At the same time, this information is logged in a corresponding file.
The actual height is measured by:
- laser sensors (Fig. 3.40)
- image processing (Fig. 3.41).

Two different laser measurement systems are currently commercially available: one is equipped with a sensor that is shifted transverse to the direction of production; the other uses several stationary sensors. Such systems are supplied by various manufacturers.

Fig. 3.41: Height measurement using image processing

b) Measurement of bulk density
As a stand-alone parameter, product height is not sufficient for the purpose of assessing quality. For instance, the properties of the hardened concrete are strongly dependent on the bulk density. For this reason, the system of block height measurement can be extended to include determination of the bulk density of the products immediately after manufacture. This requires weighing of both the empty and the full pallets (Fig. 3.42).

Empty pallets are weighed individually upstream of the machine (1). This weight is stored by the associated software and transferred along the further route of the pallet. Downstream of the production machine, each full pallet is individually re-weighed (2). The weight of the products is then determined by subtracting the weight of the empty pallet from that of the full pallet. Using the height (3) measured for selected rows of blocks, the mean bulk density of the products on the relevant pallet can be shown in a

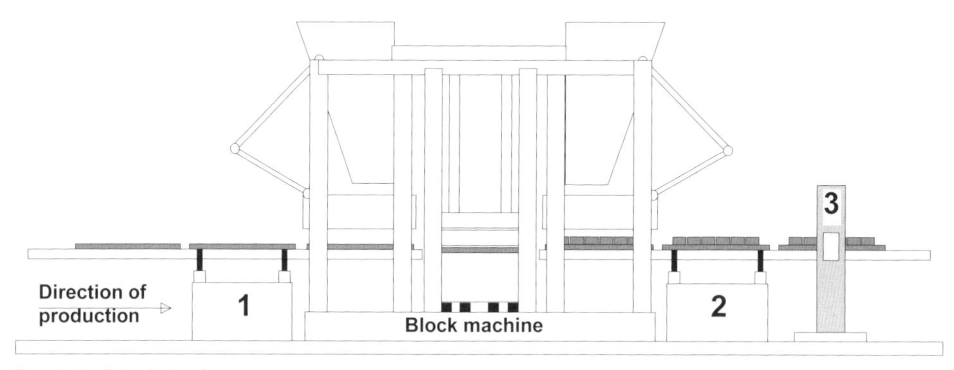

Fig. 3.42: Principle of bulk density measurement

Fig. 3.43:
System for the determination of bulk density

designated colour code and is saved in a file. Fig. 3.43 shows the technical implementation of such a system in a production line.

c) Equipment diagnostics using VibWatcher
Typical spectra of the working masses of a block machine, such as the table, mould or load, are termed spectral patterns. In [3.13], a display software (spectral pattern display) was developed that compares spectral patterns of main vibration cycles consecutively recorded during measuring runs. The "spectral pattern display" shows a comparison of the individual spectra of the same working masses that were recorded at different points in time. These spectral patterns are an appropriate means to visualise the compaction process by providing an important parameter (i.e. the acceleration of the relevant working masses) as a spectral representation immediately after the main vibration step.

This system aims to establish a basis on which to evaluate the cycles of a block machine. The display quickly informs the machine operator of all required outcomes of the respective cycle. The enormous amount of information is reduced to such an extent that the operator is able to evaluate and determine the quality of the concrete products manufactured at a glance (Fig. 3.44).

In practical operation, the frequency spectra are analysed both individually and jointly at four measuring points.

Each of the determined spectra has maximum values in certain frequency ranges. If these values lie outside these ranges, the machine setting has deviated from the reference. Fig. 3.45 shows an example of a reference spectrum. The monitoring principle is to compare a defined value with the value of the measured acceleration within selected spectral lines, which are shown in Fig. 3.45. The areas enclosed by

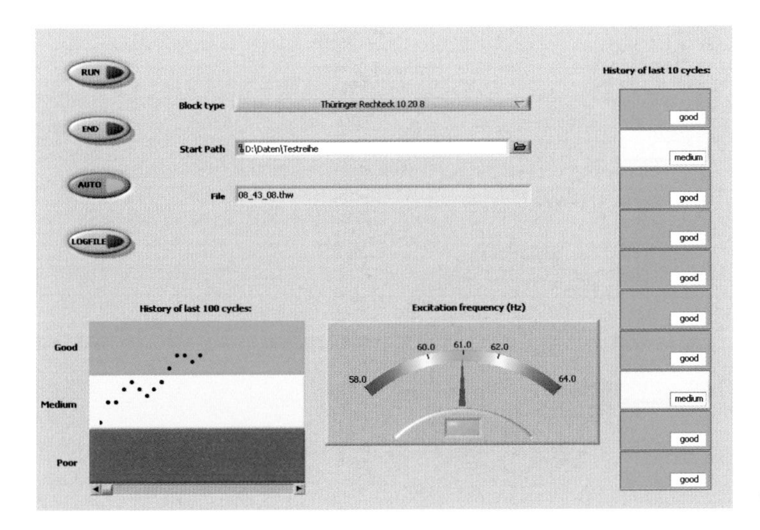

Fig. 3.44:
VibWatcher visualisation

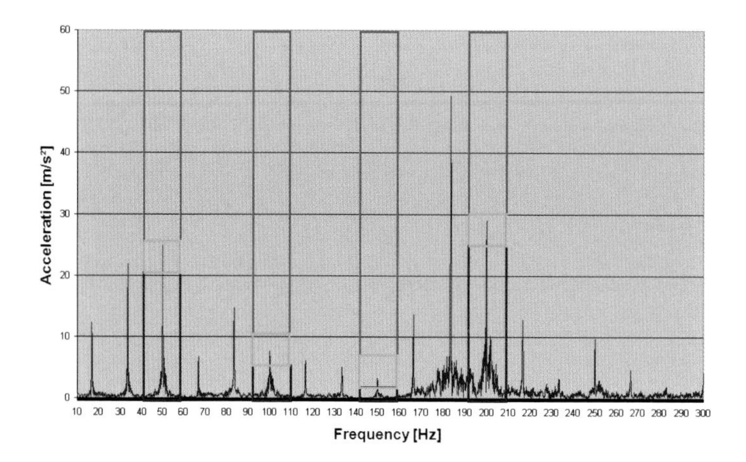

Fig. 3.45:
Spectrum within the limit

green frames represent the optimum amplitudes of these lines. The red and blue areas represent values above or below these lines. Each measured frequency spectrum is compared with this limit, which is determined only once.

This approach opens the possibility of detecting not only rapid changes caused by defects but also gradual changes due to wear, ageing (e.g. of rubber springs) and maintenance needs (spacing of knocking bars).

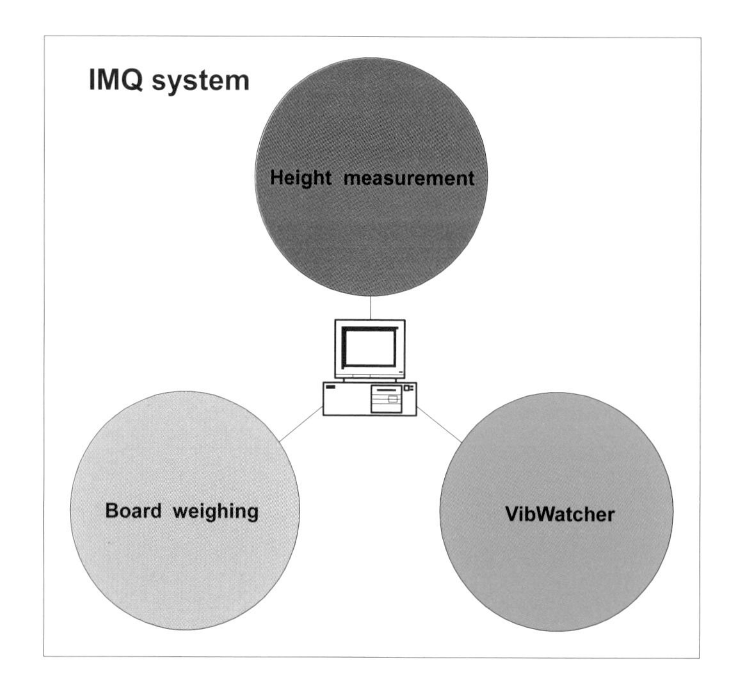

Fig. 3.46:
IMQ system

d) Intelligent monitoring for quality control: the IMQ system
The above-described online monitoring of production and equipment parameters relevant to quality is characterised by the following three autonomous systems:

- measurement of block height
- determination of bulk density
- monitoring of the compaction process

This monitoring arrangement has been merged to form an IMQ system, as shown in Fig. 3.46.

In this case, data analysis by neural networks ensures:

- early detection of quality fluctuations
- quick remediation of root causes of defects by specification of useful setting options
- information on necessary repair or maintenance work

Fig. 3.47 shows the user interface of this IMQ system.

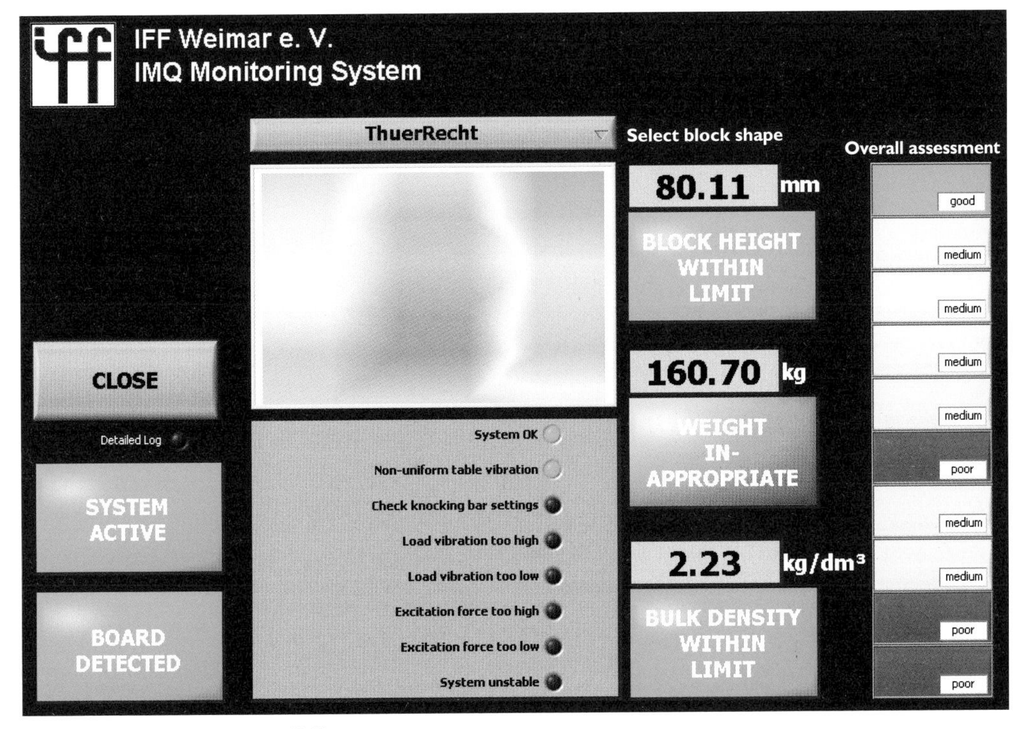

Fig. 3.47: Visualisation of the IMQ system

3.3 Egg Layers

3.3.1 Scope of Use

The term “egg layer” does not refer to the products themselves, instead it refers to the fact that these machines manufacture concrete products that are laid onto the level floor of the production line. Therefore the machine must have a mobile design to ensure continuous production. Egg layers are an economical alternative for the manufacture of concrete products whenever productivity requirements imposed on a block-making line are less stringent.

3.3.2 Configuration and Mode of Operation

An egg layer is a special type of block-making machine that is characterised by a vibration system composed of oscillating working masses, such as the table (production base), mould and load, which are used for moulding and compaction of stiff concrete mixes. The concrete products can be demoulded in fresh condition. Block-making machines can be divided into board machines and egg layers.

Whereas board machines include a base board circulation system that ensures continuous removal of finished products, the concrete blocks remain where they were produced in the case of the egg layer (Fig. 3.48), and the machine must move to the new point of production. This allows a simple design of the production process: the logistics requirements are less demanding because the base board circulation sys-

Fig. 3.48: Egg layer

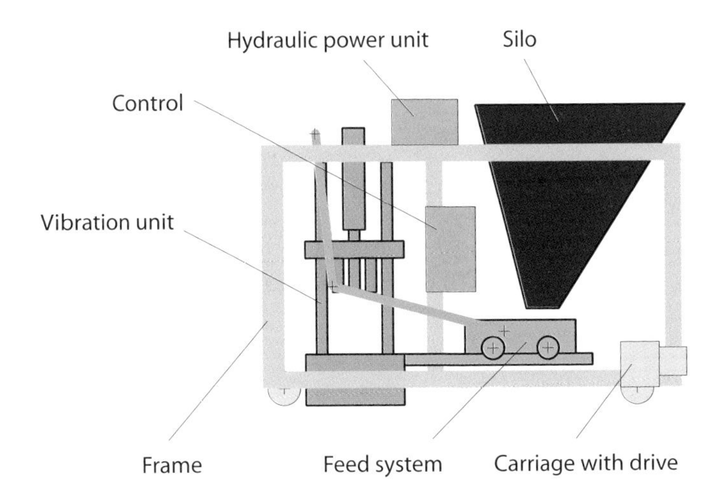

Fig. 3.49:
General configuration of an egg layer

tem and high-bay racks are no longer needed. To provide sufficient storage space nonetheless, a large, unoccupied production area must be available specifically for this purpose.

The vibration unit and the feed system are similar to their counterparts in a board machine. In its basic configuration, the egg layer includes a concrete mix silo. Face concrete may only be processed if further components are added. Fig. 3.49 shows the general configuration of an egg layer.

An inconspicuous design feature of the egg layer is its integrated travel drive. The egg layer moves on wheels guided by rails. When the end of a lane has been reached, the egg layer drives onto a below-floor carriage that travels in the transverse direction. This carriage moves to the new lane, and the egg layer drives off the carriage.

Other egg layers include a steering system to change their direction of travel without using rails. A so-called traversing car that lifts one of the travel units is used for the transfer of the machine. The frame is supported on a roller whose axis runs in a 90° offset from the egg layer's longitudinal axis. When the wheel drive is actuated, the egg layer moves about its vertical axis into the new lane.

A production cycle involves the following steps:
Concrete is conveyed from the silo into the feed box, which then moves horizontally into the vibration system and empties its contents into the mould chambers, performing shaking movements. At the same time, a pre-vibration stage can be initiated to settle the mix, thus providing more space to allow further concrete to flow from the feed box. After withdrawal of the feed box, the stamp is lowered onto the concrete mix in the mould, and the main vibration phase is initiated. The last step of the moulding and

compaction process is the demoulding phase, which is carried out by lifting the mould off the factory floor while the stamp remains in position. When the frictional force acting on the mould wall is less than the weight of the product, the stamp can be moved into its upper parking position. The egg layer then moves to its new production point. The entire sequence is repeated in the next cycle.

A complete egg laying block production line includes:

- a production area
- a machine to produce the concrete mix
- a machine to feed the concrete mix
- an egg layer
- a palletising/strapping machine
- a machine to convey the concrete block packages out of the production area
- a storage facility to accommodate the packages (alternatively, direct loading onto delivery vehicles).

Fig. 3.50 shows an example of such a production line. The size of the required production area corresponds to the daily production output plus any additional space to be kept free for technical reasons.

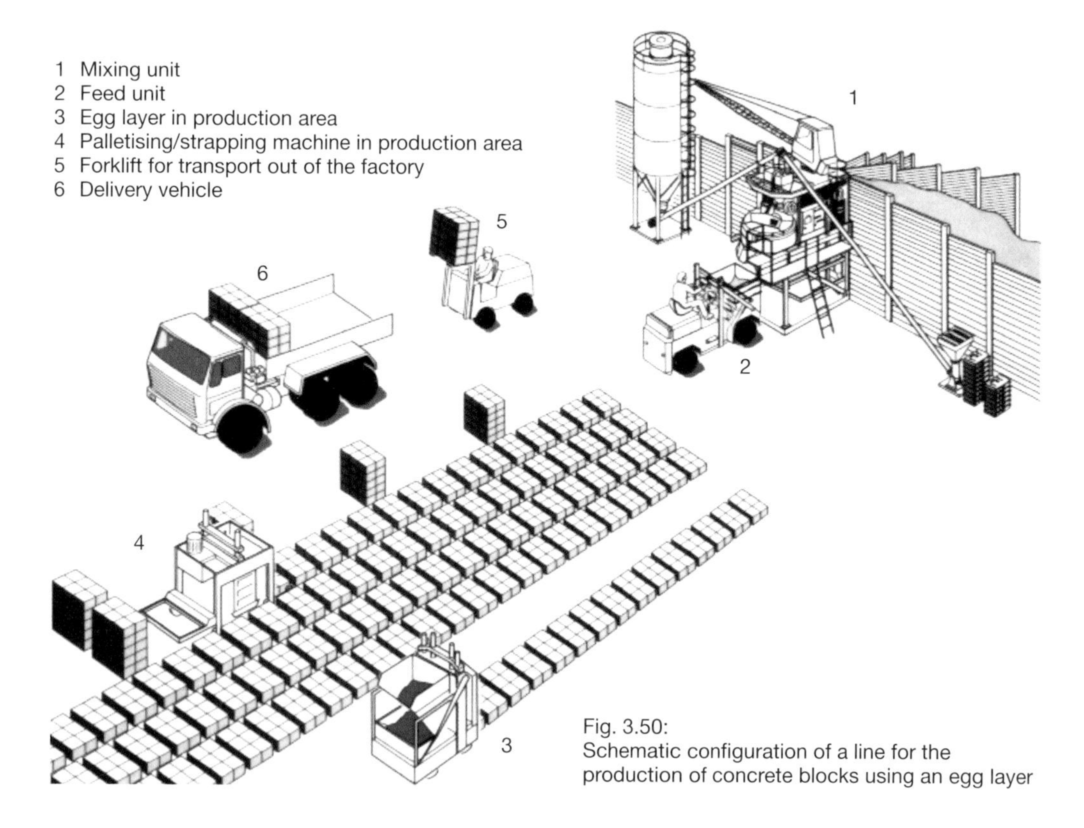

Fig. 3.50:
Schematic configuration of a line for the production of concrete blocks using an egg layer

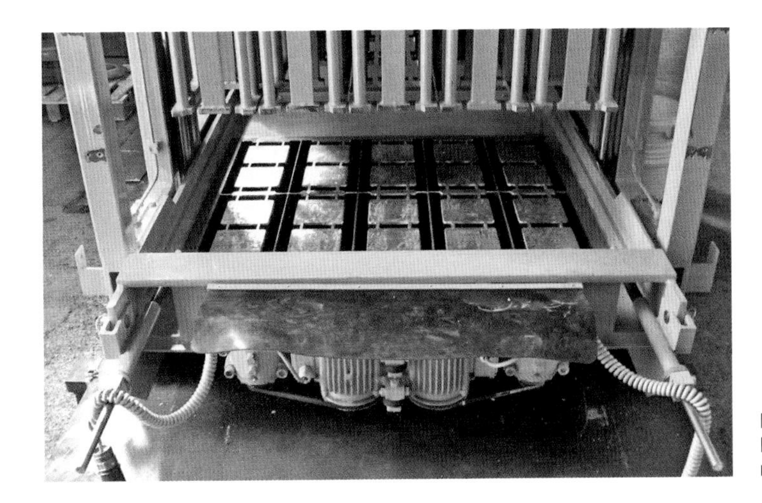

Fig. 3.51:
Egg layer with
mould vibration unit

Several functional principles apply to the manufacture of products on the factory floor; they are briefly outlined below.

Mould vibration

Mould vibration uses external vibrators fitted to the mould. No vibration occurs at the bottom of the mould. The vibrators may rotate about a vertical or horizontal axis. Additional stamp vibrators may be activated to enhance concrete compaction in the upper part of the block. Machines with mould vibration do not require a table as a means to introduce vibration. The concrete block is produced directly on the factory floor, which also simplifies machine design and shortens cycle times. However, varying floor and contact conditions may influence the vibrated compaction result (Fig. 3.51).

Table vibration

Egg layers using table vibration are the closest to the operating principle of a board machine. Products are manufactured on a vibrating table with the mould positioned on top of it. The action of the stamp is usually enhanced by stamp vibration to improve concrete compaction in the upper part of the product and its visual appearance. Following the compaction phase, the table is moved out of the production area, and the mould plus stamp are lowered to the floor.

Egg layers for products with a special shape are equipped with a mould-turning device. This system is used whenever the cross-section of the product in its manufacturing position has a greater portion of the mass in the top part and a smaller portion of the mass in the bottom, supporting position (e.g. gutters). After the moulding and compaction stage, the mould is therefore moved about its horizontal axis by 180° and lifted. The product is then demoulded onto the factory floor whilst maintaining its dimensional stability.

An egg layer is selected on the basis of the required productivity and product diversity. These are key criteria governing functionality, degree of automation, cycle time and footprint of the machine. The available options range from manually operated egg layers for pre-production or small lots to fully automatic egg layers with an additional face mix system.

Cycle time depends on the height and complexity of the product. Short cycle times for the production of hollow blocks are less than 20 seconds.

A particular issue that arises from the operation of egg layers is the level of noise generated because no soundproof cabin can be provided, unlike stationary block machines.

3.4 Slab Moulding Machines

3.4.1 Scope of Use

Slab moulding machines are used for the production of slab-shaped cement-bound items, such as paving flags. In many cases, these products consist of two layers: a coarse core mix and a finer face mix. A surface that resembles natural stone can be achieved by a special mould texture or by applying pigments to the mould.

3.4.2 Configuration and Mode of Operation

Concrete slabs are produced in a cyclic process. First, the face mix, which is made of a more flowable concrete, is fed into the mould. Following the placement of the

Fig. 3.52:
Slab moulding machine

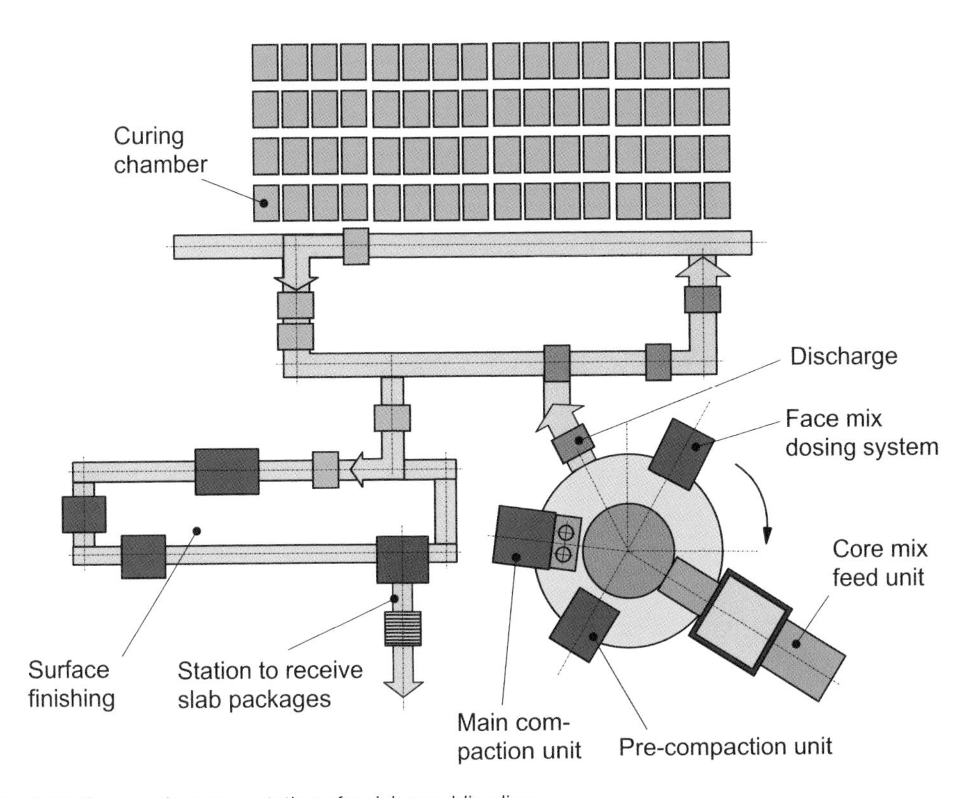

Fig. 3.53: Schematic representation of a slab moulding line

dry core mix, the slab is compacted by pressing, and moisture is transferred from the face mix to the core mix. This process creates a homogeneous bond between the two mixes.

Common arrangements of the individual units are turntable (Fig. 3.52) and sliding-bed designs. Processes were optimised to reduce cycle times to fewer than ten seconds. As a result, up to 1,000 m^2 of slabs can be produced per shift. The production process on a slab moulding machine is shown in Fig. 3.53.

Directly after moulding, the fresh concrete slabs can be finished, for example by washing out the surface. The slabs are then conveyed into the curing chamber to harden for approx. 24 hours.

The hardened slabs can be subjected to additional finishing steps. Commonly applied methods include blasting, grinding and polishing, chamfering and edging, and flame treatment. The finished slabs are then palletised and shipped.

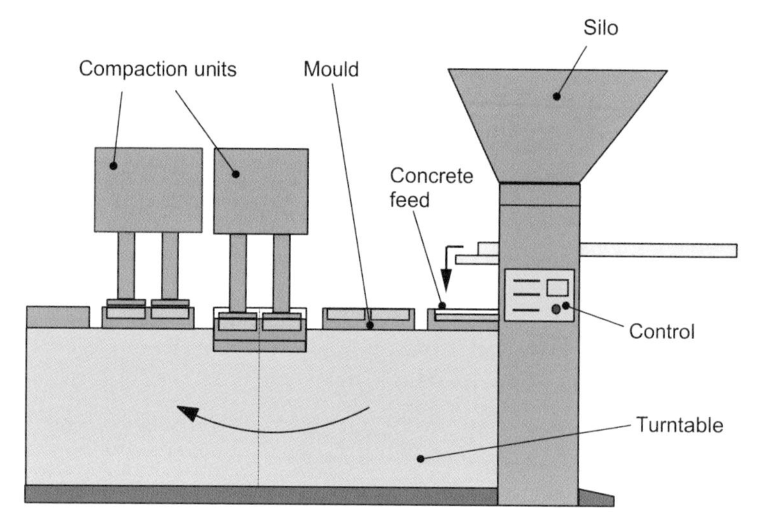

Fig. 3.54:
Basic configuration of a turntable press without upstream work steps and discharge

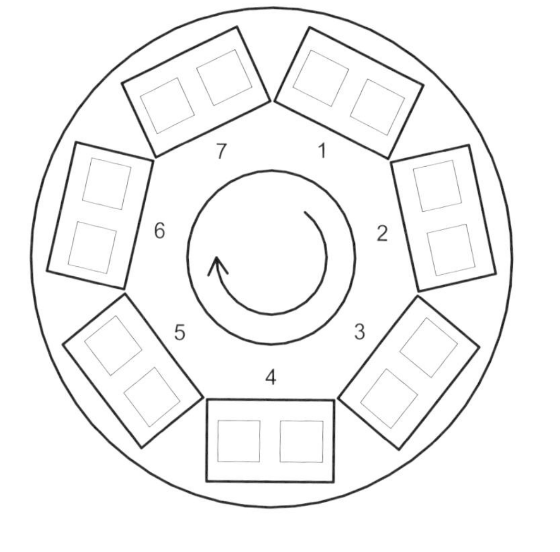

1: Prepare mould

2: Feed face mix

3: Spread face mix

4: Feed core mix

5: Pre-compaction

6: Main compaction

7: Demoulding

Fig. 3.55:
Turntable slab press (schematic view of the process flow)

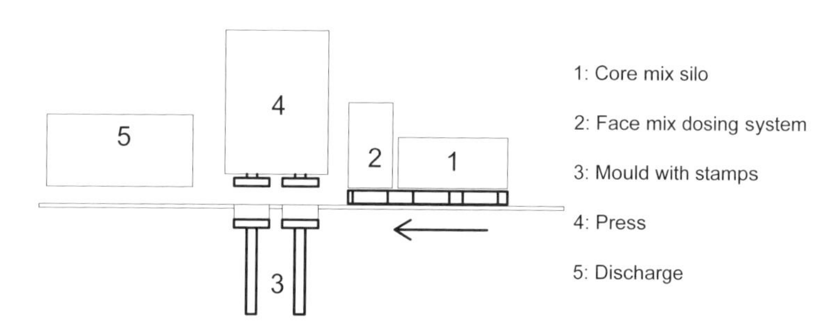

1: Core mix silo

2: Face mix dosing system

3: Mould with stamps

4: Press

5: Discharge

Fig. 3.56:
Sliding-bed slab press (schematic view of the process flow)

3.4.2.1 Turntable arrangement

In the turntable arrangement shown in Fig. 3.53, slabs are produced while the mould is rotating. The moulds arrive at the individual stations in sync with the cycle. The turntable press, also referred to as a circular press, has seven work stations (Fig. 3.55). Fig. 3.54 shows the general configuration of a turntable press.

The example of a hermetic turntable press is used to explain the functional principle. The mould is made ready at the preparation station. One of the options to achieve the intended slab design is to insert a retardant paper. At the next station, the face mix is cast into the mould from a face mix dosing system. The face mix spreader ensures a uniform distribution of the mix within the mould frame, mainly using vibration. At the fourth station, the dry core mix is cast onto the evenly spread face mix. The compaction process is carried out by hydraulic stamps and usually includes two phases. The first stage is a pre-compaction process where a significantly lower pressure is applied. The main pressing phase compacts the mix at pressures from several hundred to two thousand bars. In order to achieve a uniform pressing force in the individual mould chambers, a separate hydraulic stamp is allocated to each chamber. The slabs are taken out of the moulds after pressing and conveyed to downstream drying and finishing processes.

3.4.2.2 Sliding bed arrangement

Unlike the turntable system, the sliding bed shown in Fig. 3.56 is a stationary unit located underneath the press. The feed of core and face mixes and the demoulding process are carried out using a sliding table with suitably shaped holders. After the pressing stage, the slabs are lifted from the mould together with their stamps and moved to discharge. At the same time, concrete is poured into the next mould chambers.

3.5 Production of Concrete Roof Tiles

Concrete roof tiles are small elements used to cover pitched roofs of buildings and thus provide weather protection.

The range of standard roof tiles is complemented by custom ridge and verge tiles, ventilation tiles and other special designs. In addition to single-layer concrete roof tiles, an increasing number of products with two or more layers and special functional layers (e.g. dirt-repellent) are being developed and manufactured.

Requirements for roof tiles are described in the DIN EN 490:2006-09 (product specifications) [3.16] and DIN EN 491:2005-03 (test methods) [3.17] standards. These include requirements related to:

- water impermeability
- frost resistance
- load-bearing capacity

Table 3.6: Example of the composition of a concrete mix for roof tiles

Component	kg/m³
Natural sand up to 4 mm	1,650
Cement CEM I 52.5 R	450
Water	175
Pigments	10

Table 3.6 shows an example of a typical concrete mix for the production of roof tiles by extrusion.

The main processes used for the industrial production of concrete roof tiles are casting and extrusion; the latter is used more often.

3.5.1 Casting Process

In the casting process, the roof tiles are produced in a vertical position, standing on their side edges. The watertight, demountable moulds are arranged vertically in a battery. A special filling head is used to pump the concrete into the moulds.

Due to its workability, the concrete mix has self-ventilating properties. No additional compaction by quasi-static or dynamic forces takes place.

The concrete is cured in the mould, whose integrated thermal insulation helps to accelerate the curing process of the roof tiles.

The casting process is mainly used because it produces very smooth concrete surfaces as well as less dust and noise.

3.5.2 Extrusion Process

The extrusion process is a highly productive method of manufacturing concrete roof tiles, offering a higher production output per unit of time than the casting process.

Via an automatic dosing system, the materials are mixed according to defined weight percentages. The mix is then conveyed to the extruder, where it is extruded onto a series of metal plates, also referred to as pallets, to form a “continuous roof tile” in the horizontal direction.

This continuous roof tile is then cut to the desired size by a mobile cutting system that moves in sync with the pallet conveyor. Coloured coatings may be applied after the cutting process. The individual tiles are heat-treated in curing chambers and then palletised and stored.

Modern semi- and fully automatic extrusion systems for concrete roof tiles permit production outputs of up to 150 tiles per minute. The concrete roof tile machine is at the very heart of a roof tile production line, which includes the following components:

- mixing unit
- colour coating system
- curing chambers
- heat treatment
- demoulding unit
- palletising/packaging machines

(see Fig. 3.57).

A concrete roof tile machine usually consists of the following main components:

- concrete filling and feed unit
- tool box with spiked roller
- cutting unit with moulded blade
- conveyor system to transport the pallets

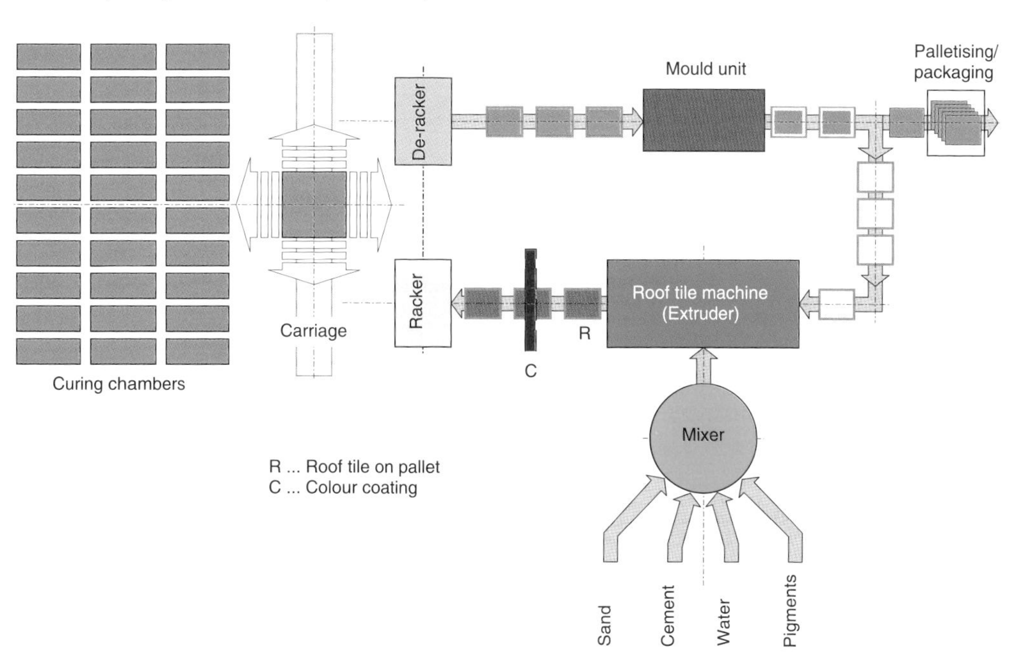

Fig. 3.57: Roof tile machine (schematic view of extrusion process)

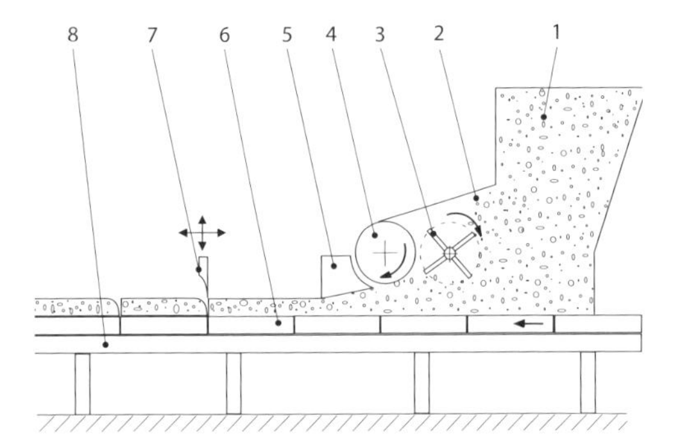

Fig. 3.58:
Basic configuration of a concrete roof tile machine
1 Concrete silo
2 Toolbox
3 Spiked roller
4 Shaping roller
5 Die
6 Pallet
7 Cutter
8 Pallet conveyor

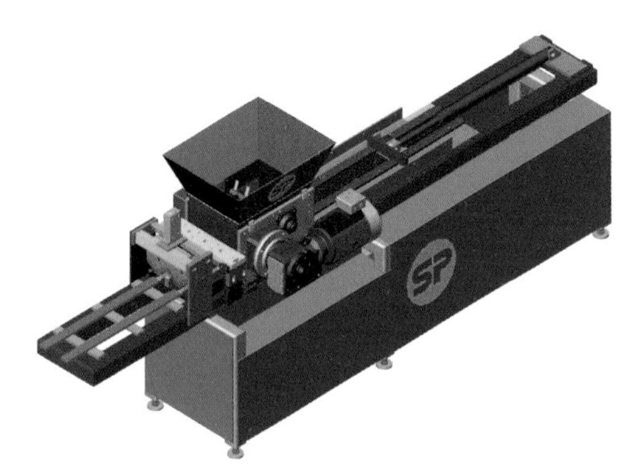

Fig. 3.59:
Concrete roof tile machine

Fig. 3.58 shows a schematic view of the basic configuration of a concrete roof tile machine. The concrete mix is fed into the concrete silo (1) from above. The downstream toolbox (2), which contains a spiked roller (3) and a shaping roller (4), ensures a uniform distribution of the concrete on the pallet (6). The main component of the concrete roof tile machine is the shaping roller (4), which shapes the top contour of the roof tile. The die (5), located at the outlet, also shapes and smoothes this contour. The feed system pushes the pallets through the roof tile machine, compacting the relatively dry concrete mix.

The continuously extruded concrete roof tile is cut to size between the pallets by the cutter (7). A blade is pressed into the extruded concrete from above. This blade has a certain shape that gives the roof tile its rounded edge on the exposed side.

3.5.3 Quality Control

To date, the finished roof tiles are usually subjected to a manual visual inspection. This type of work is not only very monotonous, the assessment is subjective and workers are prone to loss of concentration. However, modern image-processing methods can provide a seamless and objective quality control of the roof tile surfaces.

These modern testing systems are based on a computer-assisted image analysis and can be integrated into the production process at several stages. The surface of the concrete roof tile is checked for holes, cracks, bubbles and chipping. The roof tile passes through a box equipped with illumination and a camera system. High-performance, digital-output cameras with integrated memory are now available to acquire the required images. In order to identify defects or faults correctly, a uniform lighting arrangement is crucial.

Another system measures the current consumption of the roof tile machine during the moulding and compaction process: changes in the current consumption indicate changes in the production process.

3.6 Finishing and Post-treatment

Architects and clients are using concrete for an ever-increasing number of purposes. This particularly applies to cast stones made of concrete, which are regulated by DIN 18500 [3.3]. These are cement-bound products either with untreated surfaces whose visible sections receive a special pattern from the formwork, or with finished surfaces [3.13]. DIN V 18500 makes a distinction between finishing of fresh products and the subsequent treatment of hardened products.

3.6.1 Finishing of Fresh Products

The product surface may be already treated when the concrete mix is poured into the mould or formwork [3.13]. Textured inserts in the bottom of the mould structure the concrete surface, which would otherwise remain smooth (Fig. 3.60). This method can be used, for example, to create sandstone or travertine textures or floor slabs that mimic laying patterns.

However, wet-side finishing can already be carried out during the mixing process by adding pigments or creating a marbled texture. This involves adding liquid paint to the face concrete mixer and working it into the mix using an agitator. This process results in either a uniform colour or a bicoloured concrete.

Marbled concrete is produced using special metal plates located underneath the dosing nozzles of the slab moulding machine. Small quantities of the face mix pass through the slots of these plates and fall into the mould. Paint is sprayed onto the concrete mix during this dripping process. When the coloured mix hits the bottom of the mould, it spreads and creates irregular streaks.

Washing of cast stone products has become largely outdated. Following their removal from the mould, the products are freed of surface laitance, which exposes the mineral aggregates. An interesting surface texture can be achieved by selecting aggregates in suitable sizes and an appropriate intensity of the washing process.

Fig. 3.60:
Decorative finish provided by textured moulds

3.6.2 Finishing of Hardened Products

When finishing hardened products, it is crucial to adhere to the right proportions of the individual mix components in order to achieve the desired product quality. This particularly applies to the selection of aggregates, which add a great deal of the aesthetic appeal to the concrete.

This dry finishing process, i.e. an automated equivalent of stonemasonry techniques, is thus similar to the dressing of natural stone [3.13]. Blocks, slabs, façade panels and stairs are bush-hammered, pointed, stabbed, embossed, blasted, milled, ground or polished. In addition to the aesthetic appeal of the product, practical considerations become increasingly important, such as the anti-slip properties of walk-on products, easy cleaning, or strength and resistance parameters. Complete finishing lines or stand-alone machines are designed so that they can be automatically adjusted to the product being manufactured, using either a master control system or a control unit fitted locally to the machine.

Fig. 3.61:
Washing of cast stone products

Fig. 3.62:
Grinding machine

The finishing process generally starts on a calibration machine, which includes one or two work stations to equalise differences in the height of the supporting concrete layer, especially of paving blocks, using diamond grinders (Fig. 3.62). The face concrete layer can then be ground in up to ten consecutive stations.

The quality and appearance of the products to be finished determines the number and distribution of the systems for milling, smoothing and fine grinding, and whether these systems are to be fitted with diamond tools or grindstones, as well as the selection of wet or dry treatment methods.

Fine grinding is only suitable for products to be used in interiors. In contrast, products for outdoor use are subjected to rough grinding (or just milling followed by blasting) to produce the required anti-slip properties. In this process, two or more turbines are used to fire small steel balls at the product to remove cement and thus expose the aggregate.

Fig. 3.63: Curling

Fig. 3.64: Machine equipped with pointing hammers

Fig. 3.65:
Splitting of a masonry block

Curling is another finishing process in which the products are subjected to abrasion as they pass through a system of up to six carborundum-coated brushes. The technical implementation depends on the product to be treated. Depending on the composition of the face concrete layer, the products may be curled, or blasted and curled, or ground, blasted and curled. This treatment makes the surface shine whilst adding anti-slip properties.

The above types of finishing are particularly suitable for floor slabs, façade panels, paving blocks and stairs. Masonry blocks, pavers, blocks and palisade elements can also be pointed, embossed and split.

Hammers are used for pointing and embossing. These hammers are arranged in series on six to twelve pointing beams. The continuous control of the working speed and applied force makes it possible to create both fine and "antique" (roughened) exposed aggregate surfaces and/or aged products. Compared to the drum method, the aged products manufactured in this process offer the advantage of retaining the block layers and eliminating the need for re-sorting.

For masonry and hollow blocks, splitting is another option to expand the product range (Fig. 3.65).

3.7 Selection Criteria

The following table lists several key criteria to be applied to the selection of equipment and machinery for the production of small concrete products.

1. Concrete product	Which concrete products are to be manufactured?
	– shape
	– surfaces
	– colours
	
	How many shapes are to be manufactured within a single production cycle (number per production cycle)?
	
	What are the requirements for the concrete mix?
	– concrete grade
	– workability
	– mix design
	– shape of aggregates
	– maximum of aggregates
	– type of aggregates
	  – lightweight
	  – normal
	
	Properties of fresh and hardened concrete
	– density
	– strength
	– resistance
	
	Are there any specific dimensional tolerance requirements?
	
	Should face mix(es) be used?
	
	Has a finishing process been specified?
	
	Has a post-treatment process been specified?
	– of fresh concrete
	– of hardened concrete

2. Production equipment	Which type of machine can produce the required shapes? Which production process ensures that dimensional tolerances are adhered to consistently? How many products should be manufactured in which period? (multi-shift operations?) Is the intended location appropriate for the machine? (building height, ground conditions) Can the machine achieve the required production output under the prevailing conditions? Have upstream and downstream equipment and processes been designed to achieve this output? (mixer, transport out of the factory, storage facility)
3. Quality	Which parameters are key to the production process? Which parameters should be achieved by the machine? Which options exist to influence the production parameters? Which in-process quality control measures exist or need to be implemented? Which quality control measures exist for upstream and downstream processes?

4. Flexibility	How are moulds exchanged? – time – cost Which mixes should or can be processed? Which product changeovers are required, and how often? Can the production line be extended? Are breakdowns or emergencies possible? – Which? – Required actions?

4 Production of Concrete Pipes and Manholes

Concrete and reinforced concrete pipes are mainly used to construct pipelines for the supply and disposal of water. These products are thus crucial for the planning, repair and maintenance of infrastructural facilities all over the world (Fig. 4.1).

The difference between concrete and reinforced concrete pipes relates to the integration of reinforcing steel, which is carried out by inserting one or more reinforcement cages. Pipes are produced in various designs with circular, oval, or custom cross-sections, with or without base, with bell-and-spigot or rebate joints. The very popular bell-and-spigot design with a circular cross-section is shown in Fig. 4.2 together with some of the associated terms and dimensions.

4.1 Production Process

Table 4.1 shows a systematic classification of the production processes (see also [4.2], [4.8]).

Fig. 4.1: Outdoor storage facility at a concrete pipe plant

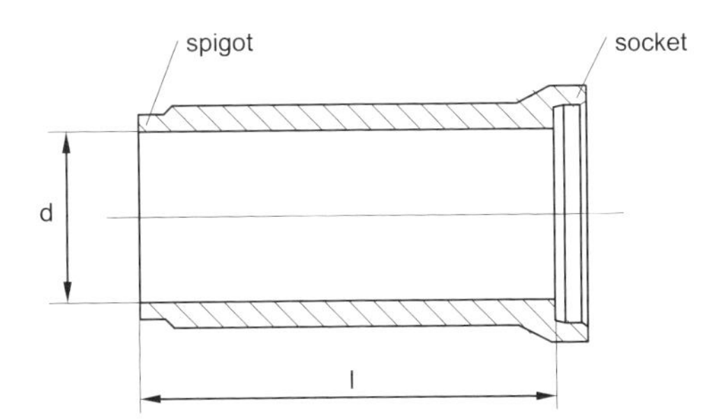

Fig. 4.2:
Circular bell-and-spigot concrete pipe, without base
d inside diameter; nominal bore DN = standardised inside diameter in mm
l installed length

Table 4.1: Production processes for the manufacture of concrete and reinforced concrete pipes

<table>
<tr><td colspan="5">Production processes for the manufacture of concrete and reinforced concrete pipes</td></tr>
<tr><td rowspan="6"></td><td colspan="4">In-mould curing</td></tr>
<tr><td colspan="2">vibration processes</td><td>horizontal processes</td><td rowspan="2">use of self-compacting concrete</td></tr>
<tr><td>in moulds using external and/or internal vibrators</td><td>on vibration tables</td><td>spinning, rolling</td></tr>
<tr><td colspan="4">Immediate demoulding</td></tr>
<tr><td rowspan="2">packer head process</td><td colspan="2">vibration process</td><td rowspan="2">combined processes</td></tr>
<tr><td>with a stationary core</td><td>with a rising core</td></tr>
</table>

The pipe machine is at the core of the pipe production line, which also includes the following components:

- concrete mixing unit
- reinforcement welding machine
- transport system for the pipes
- transport system for moulding parts (e.g. bottom ring circulation)
- pipe testing
- storage

Fig. 4.3 shows a schematic diagram of a production line.

The vibration process with in-mould curing generally uses steel moulds with corresponding demoulding options and steel rings as top and bottom pallets. The concrete mix is compacted by vibration generated by external vibrators attached to the

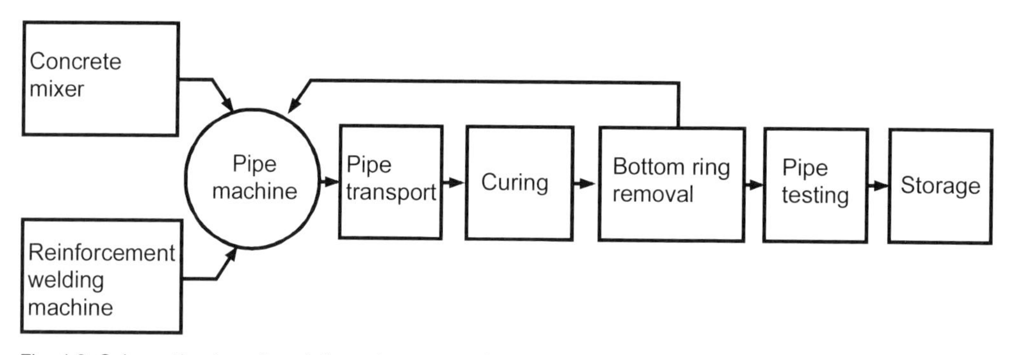

Fig. 4.3: Schematic view of a reinforced concrete pipe production line

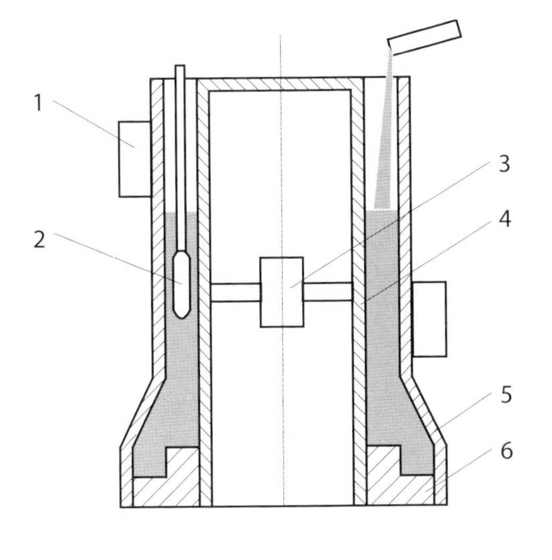

Fig. 4.4:
Working principle of in-mould vibratory compaction using mould vibrators
1 External vibrator
2 Internal vibrator
3 Core vibrator
4 Core
5 Outer mould wall
6 Base ring

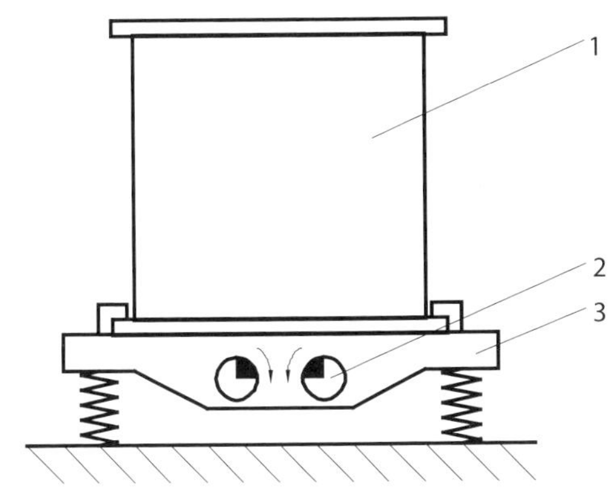

Fig. 4.5:
Working principle of compaction on a vibration table
1 Mould
2 Vibration exciter
3 Table

moulds, which are mostly flexible. For greater wall thicknesses, internal vibrators (cylinders) may also be used (Fig. 4.4).

In the compaction process on a vibrating table, the moulds are excited to vibrate on a rigid table (Fig. 4.5).

In the centrifugal spinning process, a (mostly two-part) bolted mould jacket with running surfaces rotates horizontally on a roller base. End rings are used for the socket and spigot ends. The rotational speed is so high that the poured concrete is pressed against the mould wall by the centrifugal forces (Figs. 4.6 and 4.7). After the concrete has been fed in, the speed, and thus contact pressure, can be increased for compaction purposes. The wall thickness can be varied as required by the amount of concrete fed into the mould. Different nominal bores can be produced consecutively on a single pipe machine.

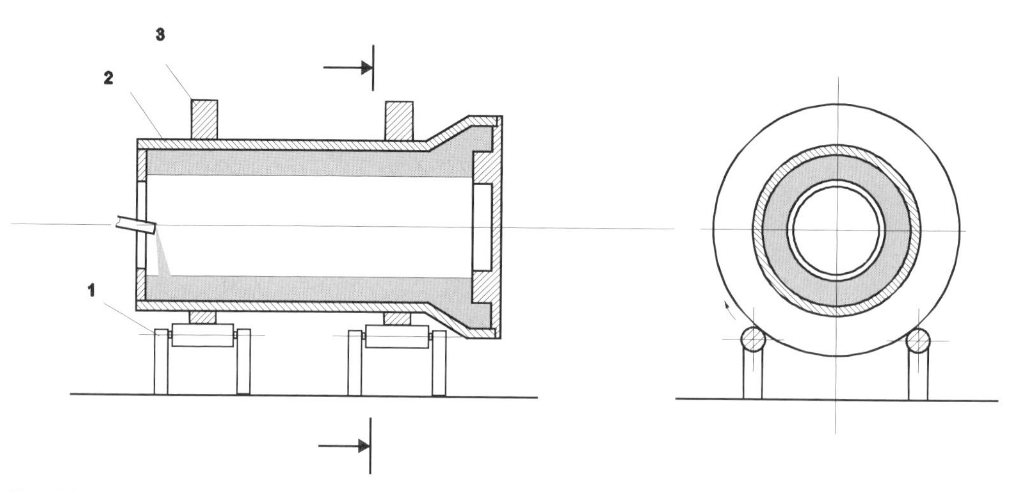

Fig. 4.6: Working principle of spinning
1 Roller base
2 Mould
3 Collar with running surface

Fig. 4.7:
Spinning machine for the production of reinforced concrete pipes

The Rocla rolling process (named after Robertson/Clarke) is similar to the spinning method. The pipe mould is attached to a roller shaft that rotates at such a high speed that the poured concrete adheres to the mould wall. Concrete is poured until the mould is no longer supported on the guide rings but rests directly on the concrete. This results in a very high degree of compaction (Fig. 4.8).

The packer head process is a vertical arrangement for the manufacture of concrete and reinforced concrete pipes with a circular cross-section. At the beginning of the production process, the pipe socket located at the bottom is compacted by external vibrators or pressing tools.

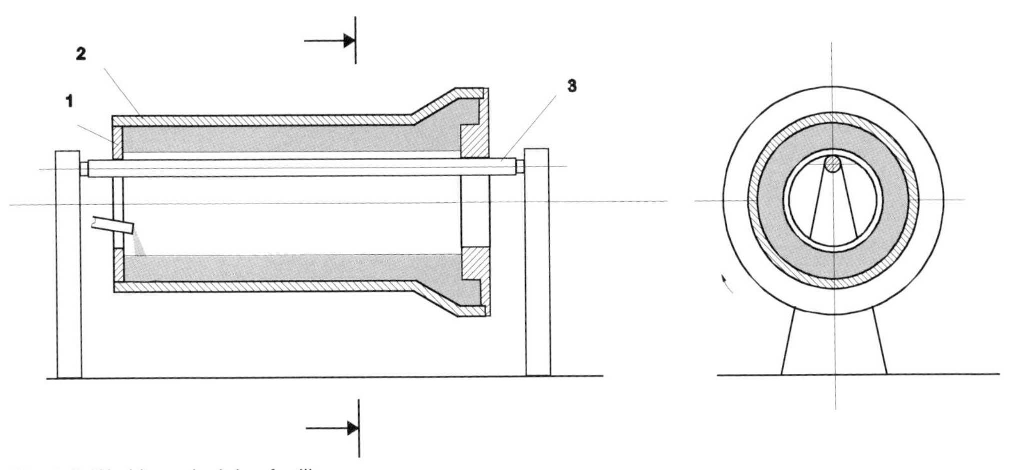

Fig. 4.8: Working principle of rolling
1 Guide rings (made up of end rings)
2 Mould
3 Roller

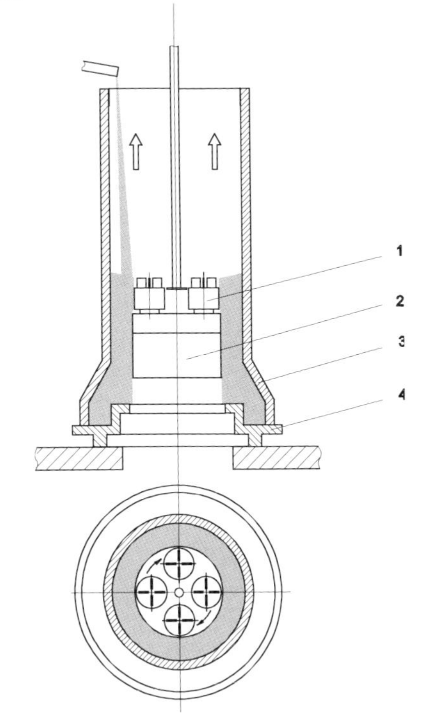

Fig. 4.9: Working principle of packer head process
1 Rollers
2 Smoothing cylinder
3 Mould jacket
4 Bottom ring

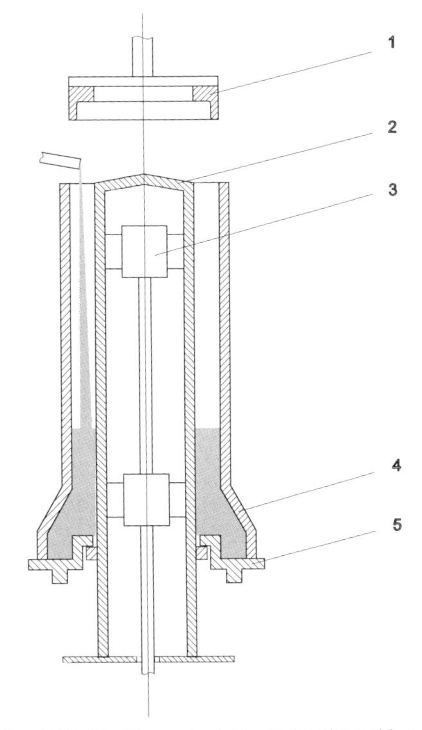

Fig. 4.10: Working principle of vibration with a stationary core
1 Spigot end shaper
2 Core
3 Core vibrator
4 Jacket
5 Bottom ring

The pipe is moulded by a rising packer head, which is composed of rotating rollers that press the concrete against the mould jacket (Fig. 4.9). An additional compaction effect is created when plates are positioned on the rollers that throw the concrete mix against the jacket. The rising speed can be controlled via the applied pressure. The roller heads counter-rotate in order to prevent twisting of the reinforcement cage. After immediate demoulding, the pipes remain standing on the bottom ring.

In the vibration process with a stationary core, concrete is poured into the space between the fixed core and the mould jacket and is then compacted by vibration (Fig. 4.10). The vibration process with a rising core differs from the stationary core process because, as the name implies, the core rises in the jacket during the pipe production process (Fig. 4.11). One of the reasons why this method has been developed relates to the uncertainties associated with the feed of concrete between the core and the jacket over the entire pipe length for reinforced concrete pipes due to smaller wall thicknesses and the integrated reinforcement cage.

Combined processes link some or all of the above functional principles to each other. For instance, vibration with a rising core is combined with other compaction mechanisms. At the top end of the core, special heads are positioned that perform a rolling,

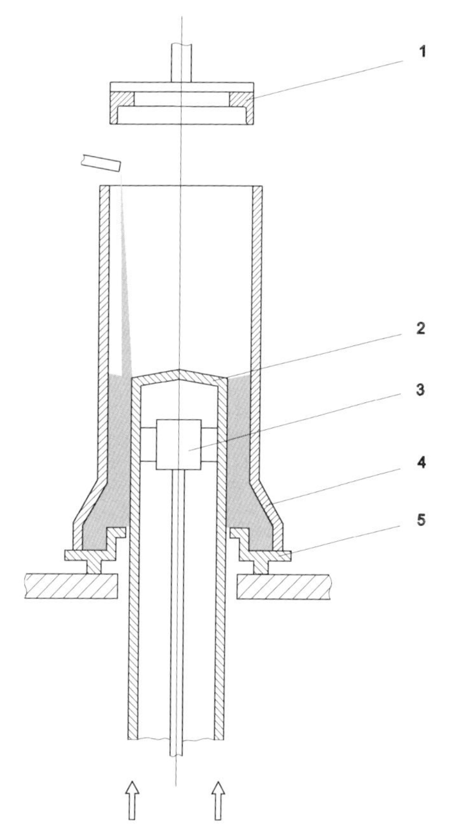

Fig. 4.11: Working principle of vibration with a rising core
1 Spigot end shaper
2 Core
3 Core vibrator
4 Outer mould wall
5 Bottom ring

Fig. 4.12:
Reinforcement cage welding machine

pressing and/or spinning process in addition to core vibration. Components that fulfil this head function include spreader rotors, packer heads or pressing tools. A combination of packer head and vibration process also creates a zone excited by vibration that propagates as the packer head is moving.

4.2 Fabrication of Reinforcement

The steel used for reinforced concrete pipes is prefabricated to form cages that consist of rings and longitudinal rebars. Rebar thickness and reinforcement ratio are adjusted to the structural requirements. A single-layer reinforcement cage is used for smaller nominal bores. Multilayer cages are used for greater pipe diameters.

Reinforcement cages are fabricated on special cage-welding machines (Fig. 4.12). Reinforcing wire is unwound from the coil, and the ring reinforcement is wound around the longitudinal bars in a helical pattern and welded at the junctions. Socket and spigot end contours can be manufactured just as conveniently as elliptical cages. The machines can be fully automated.

4.3 Pipe Machines with a Stationary Core

Fig. 4.13 shows a schematic view of a pipe machine with a stationary core. The elastically supported core (4) is held in a fixed position relative to the elastically suspended jacket (5). A bottom ring (6) is held in place by the jacket and/or by a height-adjustable core flange.

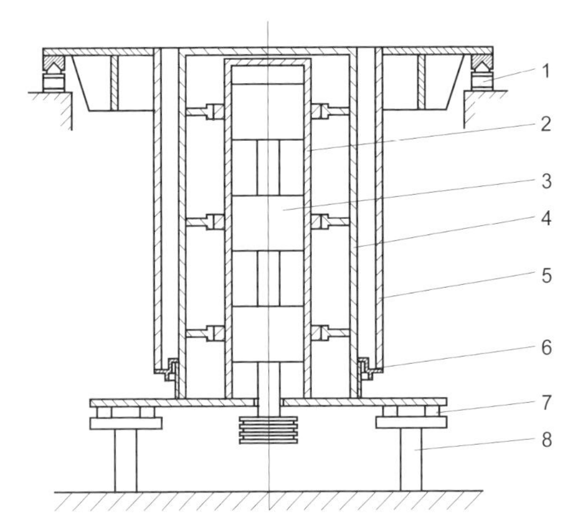

Fig. 4.13: Schematic view of a pipe machine with a stationary core
1 Jacket spring
2 Multi-stage central vibrator
3 Vibrator
4 Core
5 Mould jacket
6 Bottom ring
7 Core spring
8 Frame

The concrete mix is continuously fed into the system and progressively compacted by the core which is made to vibrate by unbalances positioned at several levels. To enable the vibration exciters to be used for different core dimensions, they are installed centrally in a multi-stage arrangement (2) within the core. When the concrete feed has been completed, a load ring moulds the spigot end in a rotary pressing process. The pipes are demoulded immediately; the sequence and demoulding movements may vary. Pipes are cured in a vertical position on the bottom ring.

Pipe machines with a stationary core are mainly used for the production of reinforced pipes with a large diameter (DN 1000 to DN 3600). Pipes with bores deviating from the standard circular cross-section can also be produced. Pipe machines with a stationary core are also used to manufacture smaller concrete pipes in production lines with

Fig. 4.14:
Production line for freshly demoulded large pipes

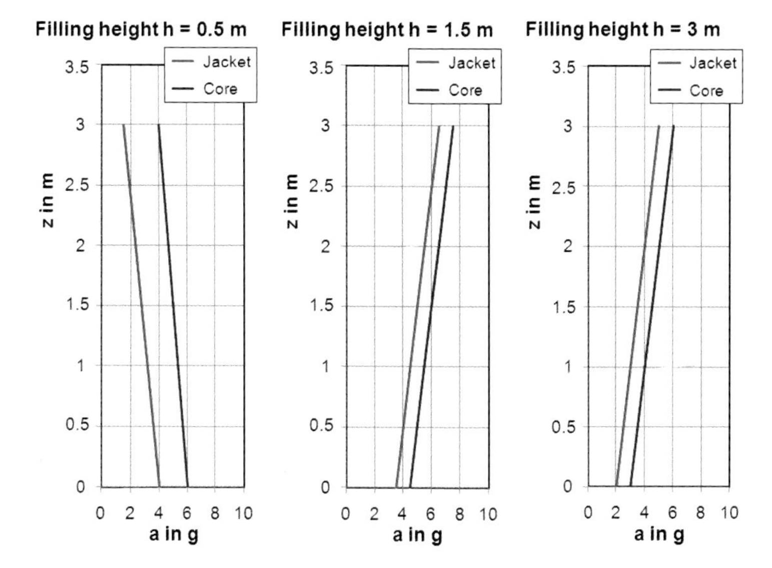

Fig. 4.15:
Types of forced vibration recommended for a pipe machine with a stationary core [4.4]

a basic configuration. This functional principle is often used in machines producing manhole rings, cones and bases.

Fig. 4.14 shows a production line for reinforced concrete pipes of larger diameters with a stationary core pipe machine installed below floor level. Such a machine may produce up to four 3-metre, DN 2000 pipes per hour. Fig. 4.15 shows a favourable distribution of acceleration parameters across a pipe machine with a stationary core for the individual production phases. Acceleration amplitudes of the core are at least 6 g at the layer being compacted. Vibrations of the jacket are slightly less, and there are relative movements between the jacket and the core. The impact of the vibrations slowly decreases in layers that have already been compacted.

4.4 Pipe Machines with a Rising Core

A schematic view of a pipe machine with a rising core is shown in Fig. 4.16. An elastically supported core is mounted on a mobile cross-beam (3). Excitation forces are applied only to the upper section of the core. The central vibrators (4) are driven via shafts or by motors mounted directly on the vibrators. During pipe production, the core is moved upwards towards the elastically suspended mould jacket (2), which creates a highly compacted zone at the core head that moves along as pipe casting progresses. The bottom ring (6) is elastically supported in a separate arrangement.

Fig. 4.17 shows a pipe machine with a rising core. The compaction zone is enclosed to mitigate noise.

Pipe machines with a rising core provide a high degree of productivity in the manufacture of concrete pipes, especially of reinforced pipes in diameters from DN 200 to DN 1600.

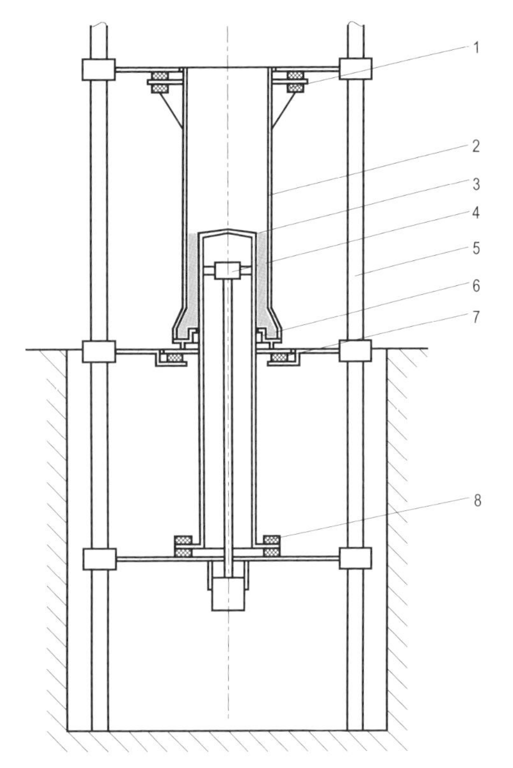

Fig. 4.16: Schematic view of a pipe machine with a rising core
1 Jacket spring
2 Mould jacket
3 Core
4 Vibrator
5 Frame
6 Bottom ring
7 Bottom ring spring
8 Core spring

Fig. 4.17:
Pipe machine with a rising core

Moulds for smaller nominal sizes are also used in a dual or triple arrangement in a single machine. A triple system offers a production output of up to sixty 2.5 m long, DN 300 pipes per hour.

Rising cores of smaller nominal sizes show a typical bending pattern under forced vibration. In an ideal case, the greatest acceleration values should be at the core head and continuously decrease towards the bottom. Bending nodes must be avoided. Fig. 4.18 shows a favourable distribution of acceleration parameters for the individual production phases. An acceleration of approx. 20 g is found at the core head. Accelerations at the jacket are lower, but amount to at least 5 g in the highly compacted zone.

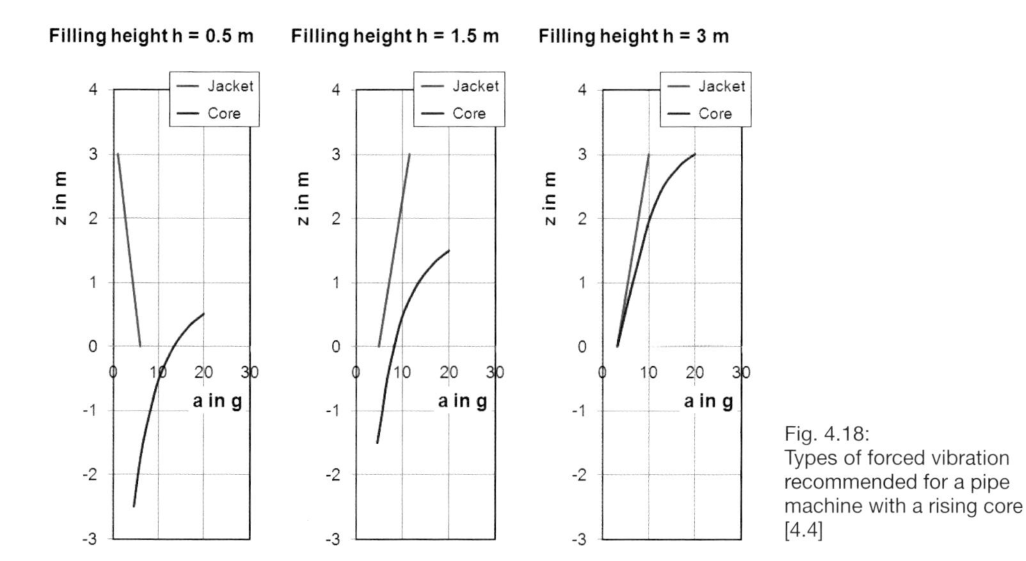

Fig. 4.18:
Types of forced vibration recommended for a pipe machine with a rising core [4.4]

4.5 Packer-head Process

A schematic view of a packer-head machine is shown in Fig. 4.19. At the beginning of a production cycle, a mould jacket (9) with bottom pallet and reinforcement cage is moved into the production area via a turntable (10). The cross-beam (1) lowers the packer head. The feed unit (4) continuously feeds the concrete mix into the jacket from above. The pipe socket is produced by a special pipe joint compactor (11), which is usually a vibration unit. The shape of the socket does not permit direct compaction of this area by the packer head. The pipe body is then moulded and compacted by the packer head, which comprises compaction rollers positioned at one or more levels (7) and a smoothing cylinder (8) (Fig. 4.20). The compacting tool is driven by a counter-rotating shaft/hollow shaft that is pulled upwards by the cross-beam. Drive motors and gears are mounted on the cross-beam. At the end of the production process, the spigot is shaped at the height of the feed cross-beam (5). Once the packer head has been moved out of the pipe area, the turntable rotates the mould jacket with the finished pipe out of the line and simultaneously moves a new jacket into the production position.

A forklift, crane or handling robot transports the pipe plus jacket to the demoulding station and pulls off the jacket so that the freshly demoulded pipe remains in vertical position on the bottom pallet. The jacket is fitted with a new bottom pallet and reinforcement and is placed onto the turntable for the next cycle.

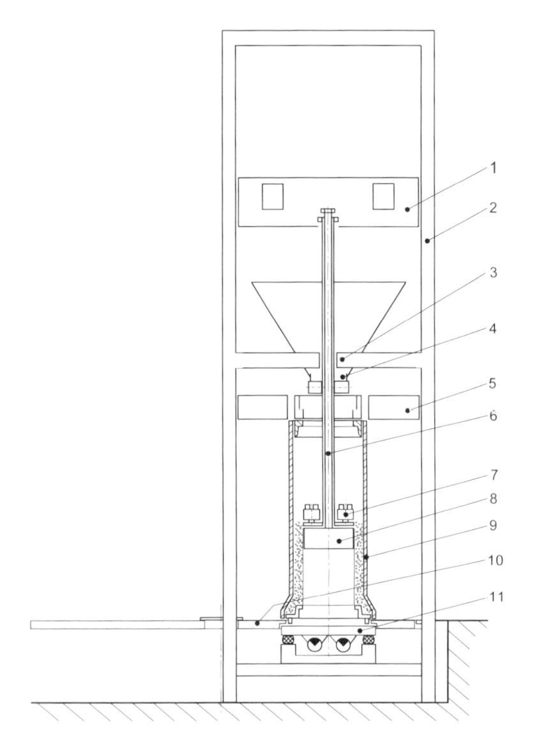

Fig. 4.19: Schematic view of a packer head pipe machine
1 Cross-beam
2 Machine frame
3 Intermediate bearing of shaft
4 Feed unit
5 Feed cross-beam
6 Shaft/hollow shaft
7 Compacting rollers
8 Smoothing cylinder
9 Mould jacket
10 Turntable
11 Pipe joint compactor

Fig. 4.20:
Packer head

Packer-head pipe machines are used to produce circular concrete and reinforced concrete pipes with nominal bores from DN 250 to DN 2000. Up to thirty DN 300 pipes of 2.5 m length can be produced per hour.

Pipes manufactured on a packer-head pipe machine have a very smooth inner surface. The external surfaces are slightly rougher. One of the worst production faults is twisting of the reinforcement cage due to inadequate matching of the individual processes. In such a case, the rotating tool triggers torsion of the reinforcement cage during the production process. After demoulding, the cage attempts to move back into its original position, which may create reinforcement shadow lines in the concrete or even completely destroy the freshly demoulded pipe. This torsional moment acting on the reinforcement cage can be compensated by counter-rotation of the smoothing cylinder and the rollers. Drives with a separate speed control help to match the individual processes. Furthermore, the transition from the pipe socket to the body needs to be monitored particularly closely during the process because different compaction methods have been applied.

The advantage of the packer-head process is reduced noise. However, this positive characteristic is sometimes compromised by the very high noise levels associated with compaction of the socket.

A wide variety of options exist for the design of the compaction tools, including the number of roller levels, tool diameter increments and integrated fittings for spreading concrete [4.1]. There is also a version with driven rollers.

The concrete mix must be designed to suit the production process. It requires a stiff concrete with sufficient green strength that compacts to a dense microstructure and does not stick to the tools. Table 4.2 shows an example of a concrete mix suitable for the packer-head process.

One of the options developed for packer-head machines and pipe machines with a rising core is the production of double-layer pipes. An inner layer made of a special material that is created directly during the production process further increases the corrosion resistance of concrete and reinforced concrete pipes [4.6].

Table 4.2: Example of a concrete mix for the packer-head process

Constituent	Unit	Proportion
Sand 0–4	kg/m³	900
Gravel 4–8	kg/m³	200
Chippings 5–8	kg/m³	850
CEM I 42.5 R	kg/m³	320
Fly ash	kg/m³	60
w/c	-	0.38

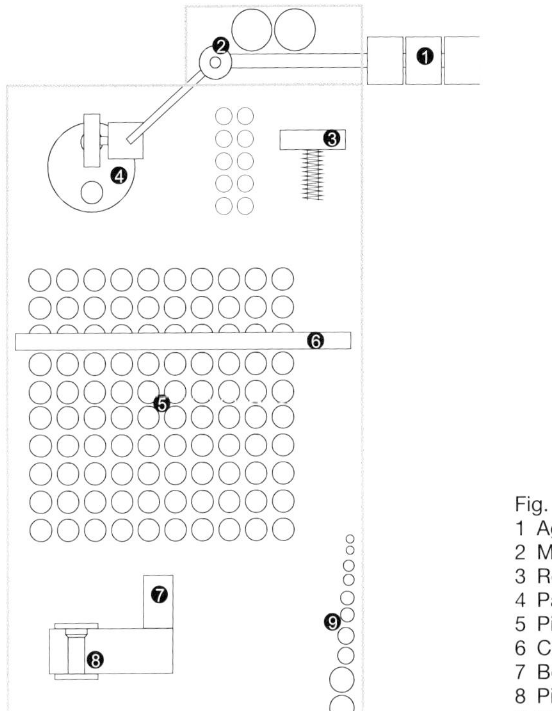

Fig. 4.21: Schematic view of a pipe production line
1 Aggregate silo
2 Mixer
3 Reinforcement welding machine
4 Packer-head pipe machine
5 Pipe storage
6 Crane
7 Bottom pallet removal
8 Pipe test rig
9 Mould storage

Fig. 4.21 shows a schematic view of a layout for a pipe production line with a packer-head machine.

4.6 Wet-cast Process

The production of pipes in vibration moulds with in-mould curing is a tried and tested process that is characterised by a high surface quality and dimensional accuracy of the finished products as well as reproducibly consistent processing of low-cost concrete mixes. This process is used, for example, to produce large jacking pipes, custom cross-sections and manhole components. Elements exceeding a length of six metres and a weight of 30 tonnes can be produced. The mould is generally used for only one pipe per day because of the in-mould curing process. Two pipes per day can be produced by taking special concreting measures and heat-treating the products to accelerate curing.

Fig. 4.22 shows a large pipe mould equipped with external vibrators for the wet-cast process. The flexible external mould is braced in the circumferential and longitudinal directions and carries external vibrators that are distributed across its circumference and height. The internal surfaces of the pipe are shaped by a core mould. A bottom

Fig. 4.22:
Large pipe mould equipped with external vibrators

ring that provides the bottom contour of the mould is positioned between the core and the external mould.

After the mould has been prepared and fitted with the reinforcement cage and any embedded parts, it is continuously filled with plastic to flowable concrete, which is then compacted. This process takes approx. 10 to 15 minutes for a 3 m long DN 2000 pipe. Curing takes place in the mould. Mould core and jacket are removed when the concrete has developed a strength sufficient for demoulding. To facilitate this process step, the core is usually fitted with a wedge-shaped device that reduces its circumference. The external mould is either expandable or has a multi-part design to assist demoulding.

To manufacture high-quality products, vibration must be introduced into the concrete so that vibrations are of a sufficient magnitude, an appropriate type and with a uniform distribution (see compaction parameters in Section 1.1.5.2).

Unlike flat moulds, round moulds present particular challenges, such as marked variances in rigidity in the tangential and axial directions. The characteristics of continuously filled vertical moulds are also subject to continuous alterations as the filling level changes. Fig. 4.23 shows an example of the vibration characteristics of a large pipe mould determined by an FEM model.

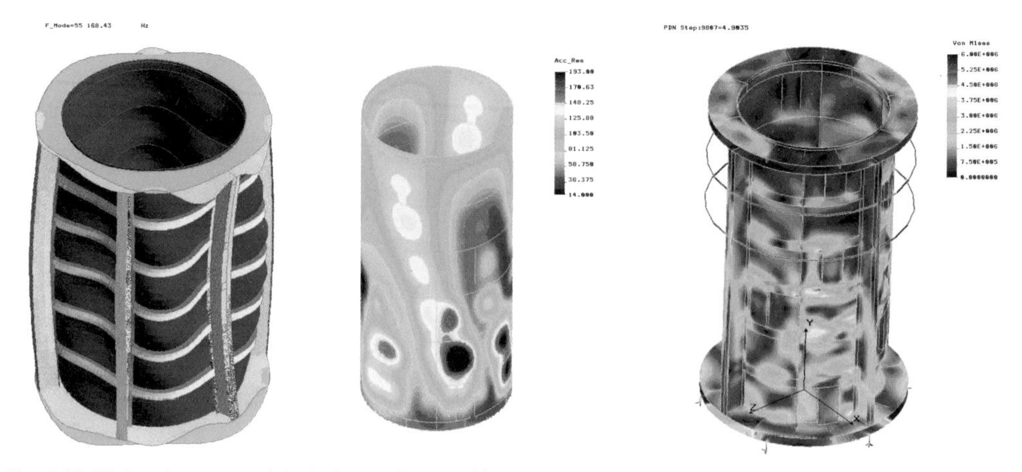

Fig. 4.23: Finite-element model of a large pipe mould
left: representation of a modal component
centre: distribution of acceleration for forced vibration
right: distribution of stresses

Like all flexible vibration moulds, the external mould has several natural frequencies that lie in the excitation frequency range and shows changes in the distribution of the vibration parameters in the case of forced vibration. Fig. 4.24 shows a measured example to illustrate the influence of the excitation frequency. Two points of resonance are clearly visible at approx. 125 Hz and 165 Hz.

For the uniform introduction of vibration, key parameters such as mould design, vibrator positions, excitation frequency and phase position must be matched to each other. The

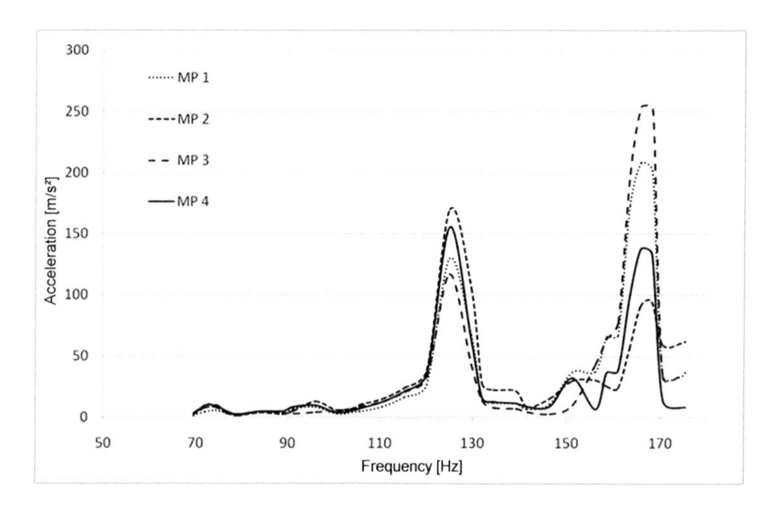

Fig. 4.24:
Amplitudes of acceleration measured at a large pipe mould depending on the excitation frequency at several measuring points (MP)

analysis and evaluation of the stresses occurring in the material in relation to material strength permits conclusions as to the durability of the equipment. Vibration concepts to be considered include, among others, beat vibration, multi-frequency excitation and variations in the excitation frequencies. When implementing these concepts, the external vibrators and their control are also important, along with the vibration-related characteristics of the moulds [4.7].

For vibration moulds, external vibrators with an asynchronous motor and unbalance excitation are used almost exclusively. Compared to other types of drives, such as universal motors or compressed-air units, asynchronous motors have a robust design and a relatively steep characteristic curve in their operating range. The latter is required for accurate control of the excitation frequencies in order to utilise the previously determined vibration-related properties of the pipe moulds.

For vibration moulds, excitation frequencies above the grid frequency of 50 Hz are generally appropriate, and frequencies around 100 Hz are normally used. As the maximum rotary frequency of an asynchronous motor cannot be higher than the supply voltage frequency, frequency converters that generate a sufficiently high electrical frequency at the output are normally used.

The development of low-cost power semiconductors and digital controllers has enabled the use of several independently operating frequency converters that can be automated with precise control of the excitation frequencies. These options are used to accurately match the vibration equipment to the moulds.

In general, the following operating modes of asynchronous external vibrators can be used to influence the introduction of vibrations into the concrete mix:

- Self-synchronisation
  If one or more external vibrators are relatively closely coupled mechanically, this will result in a synchronous operation in which the relative phase position of the unbalance exciters is constant, i.e. the external vibrators are running at the same speed. A mechanical coupling occurs if the vibrations of an external vibrator have an effect on the operation of another external vibrator. The advantage of self-synchronisation is its relatively strong and defined excitation of the vibration mould. However, in reality, a steady state will usually occur in which not all sections of the mould may be sufficiently excited.
- Beat vibration
  Beat vibration occurs in the absence of any strong mechanical coupling of the external vibrators. This results in slightly different speeds whose superposition leads to rising and falling vibration amplitudes. Due to the continuously changing relative phase position of the unbalance exciters, the vibration mould is constantly being subjected to new vibration states, which results in a more uniform introduction of

vibration over the compaction period. On the other hand, the operating noise that increases and decreases with the vibration amplitude is often considered unpleasant.

- Specific variation of the excitation frequencies
  As in beat vibration, this mode aims to generate a vibration mould excitation that is as varied as possible. This happens by adjusting and re-setting the excitation frequencies during the compaction process. This can either be simple switching from one frequency to the next or continuous changes through the frequency ranges.
- Specific excitation of modal components
  Provided the modal components of a vibration mould are known, these can be excited in a specific manner via the excitation frequency controller. This provides the opportunity to use cost-efficient vibration equipment to introduce a sufficiently high degree of vibration into the concrete mix even in relatively rigid moulds or to improve compaction with an existing piece of equipment. Similar to the synchronous operation of the unbalance exciters referred to above, this mode may lead to steady states and non-uniform introduction of vibration. If the modal components are known, however, these phenomena can be avoided by judicious control of the external vibrators.

Alternatives to compaction by external vibrators include internal vibrators, vibrating tables or self-compacting concrete.

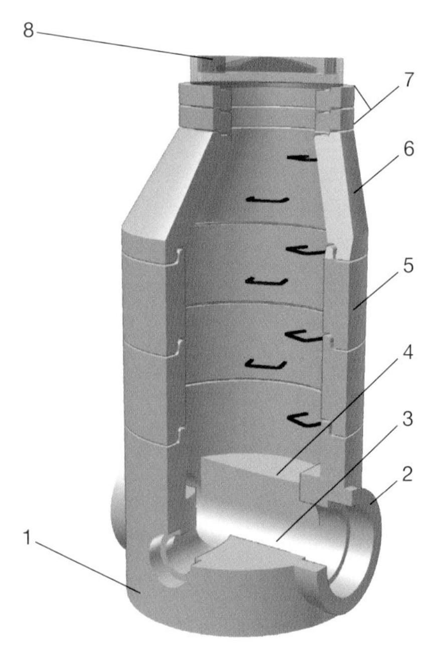

Fig. 4.25: Example of a manhole assembly
1 Manhole base
2 Connector
3 Channel
4 Platform
5 Manhole ring
6 Manhole cone
7 Top ring
8 Manhole cover

Fig. 4.26:
Manhole ring machine

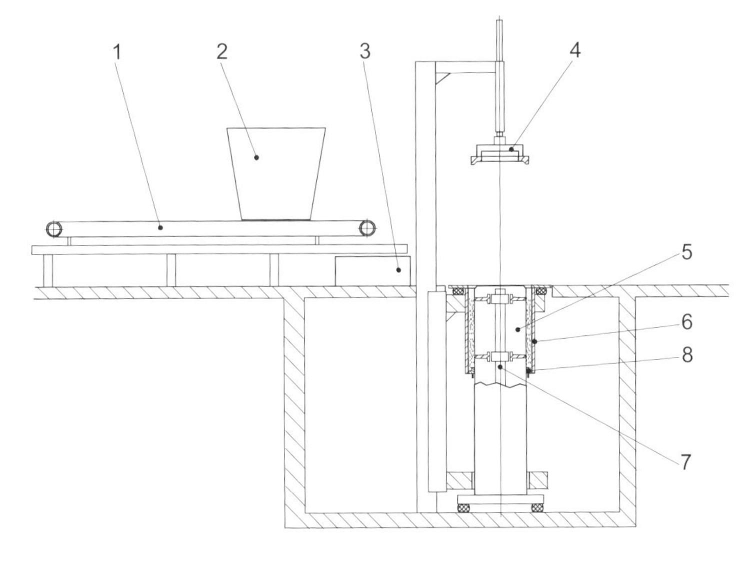

Fig. 4.27: Schematic view of a manhole ring machine
1 Conveyor
2 Silo
3 Spreader
4 Spigot end shaper
5 Mould core
6 Mould jacket
7 Central vibrator
8 Bottom pallet

4.7 Production of Manhole Rings and Bases

In sewer systems, vertical manhole units are linked to the horizontal pipelines for the purposes of access and inspection. Fig. 4.25 shows a typical manhole assembly made of concrete components. All of these products can be prefabricated at a precast plant. Typical products are manhole rings, bases and cones.

Large volumes of manhole rings are produced on manhole ring machines (Fig. 4.26) using stiff concrete mixes. Items are demoulded while the concrete is still fresh.

Fig. 4.28:
Special core for the insertion of step irons

The basic configuration of a manhole ring machine is equivalent to a pipe machine with a stationary core. Fig. 4.27 shows a schematic view of a manhole ring machine.

Insertion of the bottom ring, positioning of the jacket relative to the core, filling and demoulding are largely performed by machines and can be fully automated.
A particular challenge in the production of manhole rings is the incorporation of step irons. Specially designed shield-type cores (Fig. 4.28) are used into which the step irons can be inserted during the production process and thus fixed during moulding and compaction of the concrete mix. During demoulding, the shield is retracted so that it releases the step irons. When designing such cores, the associated reduction of the core rigidity caused by the cut-outs in the shield and the vibration resistance of the shield mechanism need to be monitored particularly closely. The impact of vibrations on the shield may differ significantly from the other areas of the core where vibrations are introduced, which may lead to variances in compaction in this zone.

A particularly challenging task in the production of manhole bases is shaping of the channel, which usually requires custom moulding owing to the different nominal bores and number of pipes to be connected, as well as differing angles and gradients.

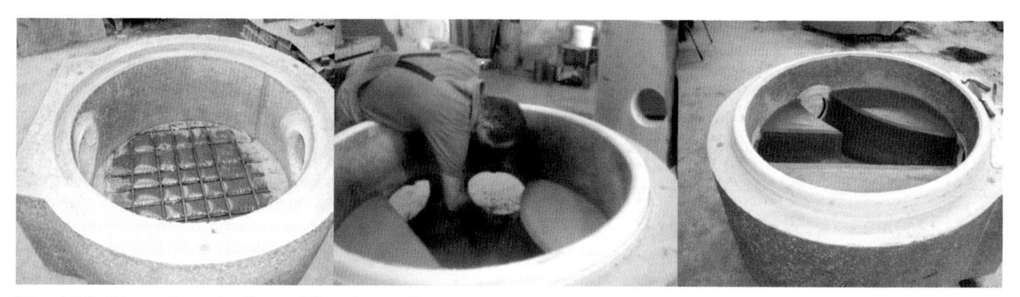

Fig. 4.29: Manual production of the channel
left: rough casting
centre: manual moulding of the channel
right: finished manhole base

Fig. 4.30:
Plastic channel

Manhole bases are usually produced upside-down. Steel moulds are available for frequently manufactured channel designs, such as straight and 90° inlets. These moulds are positioned on the core head in manhole ring machines or used in a wet-cast process.

The moulding of customised channels usually involves extensive manual work. The manufacture of manhole bases starts with the production of an external part (rough casting; see Fig. 4.29) on appropriate compaction systems. The channel is then added in a manual process. In this step, tamped concrete is fed into the rough casting and the channel is shaped by hand.

This production method has certain disadvantages such as the lower durability of the manually placed concrete, the relatively large number of employees needed and the associated high labour cost, as well as in terms of health and safety because of the considerable physical strain on the employees producing the manhole.

Another method of producing manhole bases involves the use of prefabricated plastic mould components (see Fig. 4.30). These parts are usually fabricated individually elsewhere. Use of these moulding components enables machine-based production of manhole bases.

In this process, the plastic mould remains in the manhole base as permanent formwork. Areas where the manhole makes contact with the wastewater consist of PP (polypropylene) or GRP (glass-fibre reinforced plastic). This design makes the manhole base very durable. Disadvantages of this process are not only the high cost and the amount of energy required for pre-fabrication of the plastic moulds, but also the fact

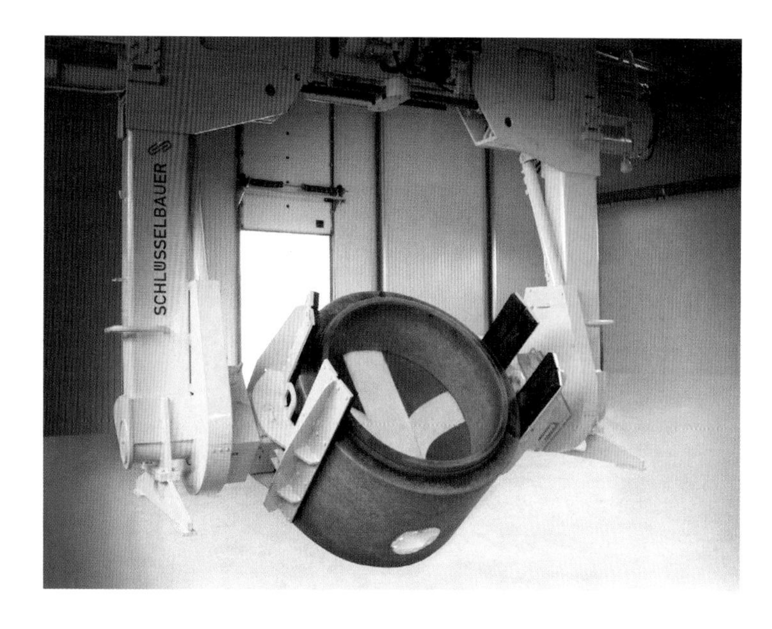

Fig. 4.31:
Concrete manhole base being turned into its installation position with the plastic moulds still attached

that these parts are produced by an external supplier, which results in a longer lead time. Consequently, the response to short-notice customer requests is also longer.

Another method uses custom-cut moulding parts made of rigid polystyrene foam (Fig. 4.31). The permanent formwork parts are cut to size using a hot-wire device and glued together to produce a standard range of moulding parts. These parts are placed on base cores together with the pipe connections that are tailored to the various pipe designs, and the concrete is poured. After demoulding, the parts can be shredded and recycled.

4.8 Curing and Pipe Testing

After the pipes have been demoulded, the concrete must be protected against drying because cement requires water for hydration. Appropriate protective measures are:

- enveloping the products with plastic (Fig. 4.32)
- water spraying
- coating with protective films (curing agents) and
- transport of products into curing chambers, particularly in pallet circulation systems

When the pipes have reached a sufficiently high early strength, they are finished and tested (Fig. 4.33).

Special equipment is used to remove and clean the bottom rings. In some cases, the spigots are milled to ensure an exact fit of the pipe connection. Pipes are tested

Fig. 4.32:
Curing of concrete pipes underneath tarpaulins

Fig. 4.33:
Pipe test rig

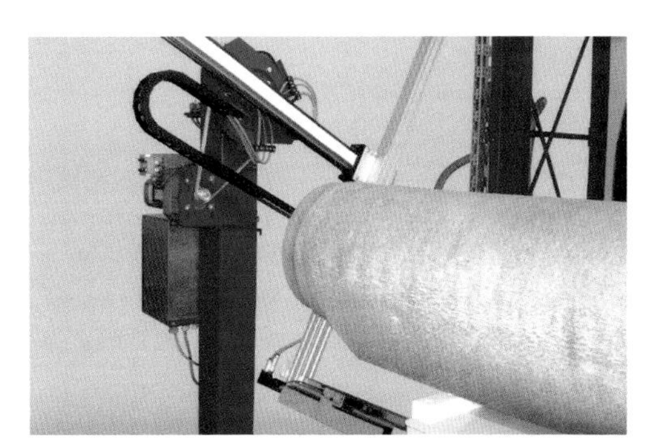

Fig. 4.34:
Measuring device for pipes

Fig. 4.35:
Leak testing of pipe walls

for their dimensional accuracy and tightness during the production process. Fig. 4.34 shows a device for measuring spigots. A leak testing rig is shown in Fig. 4.35. End covers provide a tight closure of the pipes, and water or air is used to test the impermeability of the walls.

4.9 Quality Control, Characteristics of Defects

4.9.1 Typical Pipe Defects and their Causes

4.9.1.1 Degree of compaction

A general compaction defect exists if the compressive strength of the samples taken from produced pipes is significantly lower than the compressive strength parameters measured with test cubes produced from the same concrete mix in the laboratory, taking account of the concrete technology guidelines pertaining to the influence of shape of the test specimens.

This phenomenon can be explained by assuming a combined effect of parameters related to materials, processes and equipment on the concrete quality. Test cubes are generally produced on laboratory-scale vibrating tables using different excitation frequencies, acceleration parameters, compaction periods and other boundary conditions. In some cases, impact-like actions result from 'rapping' of the cube mould on the vibrating table. The compressive strengths of these test cubes can only be achieved in the actual pipe production process if the degree of concrete compaction is uniformly high.

The degree of compaction k_V is the percentage ratio of the bulk density of the fresh concrete ρ_{fr} achieved during the actual compaction process to the fresh concrete bulk density $\rho_{fr,0}$ that is theoretically possible according to the material volume calculation (Equation 4.1):

$$k_v = \frac{\rho_{fr}}{\rho_{fr,0}} \cdot 100\% \qquad (4.1)$$

There is a clear correlation between the compressive strength of the fresh concrete and the degree of compaction: the compressive strength sinks as the degree of compaction decreases.

Fig. 4.36 shows a core sample taken from a large pipe. The degree of compaction was obviously insufficient, which is why ways to improve the moulding and compac-

Fig. 4.36:
Core sample of a large pipe in the original condition

Fig. 4.37:
Core sample after modification of the key parameters

tion process were sought. Its compressive strength amounted to only 20 N/mm² whereas the compressive strength of the test specimen produced from the same mix was 50 N/mm².

Fig. 4.37 shows a pipe core sample that was produced with modified equipment settings and thus altered key parameters. The degree of compaction was significantly improved by increasing the acceleration amplitudes and extending the compaction period.

4.9.1.2 Local compaction defects

One of the problems seen in production is the occurrence of local compaction defects in pipes (Fig. 4.38). Typical faults include defects on the front and rear faces of the pipe and poorly compacted sections along the length (Fig. 4.39).

Fig. 4.38:
Local compaction defect

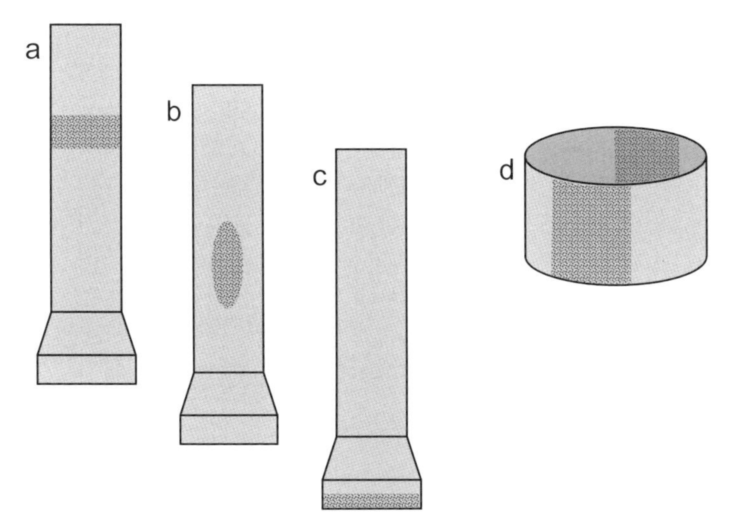

Fig. 4.39:
Schematic view of various local compaction defects
a) poorly compacted section
b) defective spot
c) compaction defects at the pipe joint
d) different degrees of compaction around the pipe surface

Such defects are caused by non-uniform vibration. One of the key requirements for compaction equipment is that vibrations are introduced into the concrete not only to a sufficient magnitude and at an appropriate frequency but also with a uniform distribution pattern. Another factor to be considered in pipe production is the progress of pipe casting during the manufacturing process. Local compaction defects may occur due to a variety of causes.

Compaction variances around the circumference of the pipe may be associated with similar differences in the distribution of the acceleration parameters, which result from directional variances in the applied excitation force, jacket and core support and/or component design. Shield-type cores used in manhole ring machines are a typical example. The shield used to incorporate the step irons may show a vibrational behaviour that differs significantly from that of the core. In addition, the shield cut-outs compromise the rigidity of the core structure so that acceleration variances may occur around its circumference.

Differences in distribution across the surface of the mould jacket occur, for instance, if one of the natural bending frequencies of the jacket is close to the excitation frequency [4.3]. This may be the case for mould jackets with a large diameter or with a low degree of circumferential bracing.

Small areas with poor or excessive compaction may be due to bulge-shaped natural vibrations of the surface plates of the jacket or core. Local disturbances such as transport anchor fastenings also contribute to this type of defect.

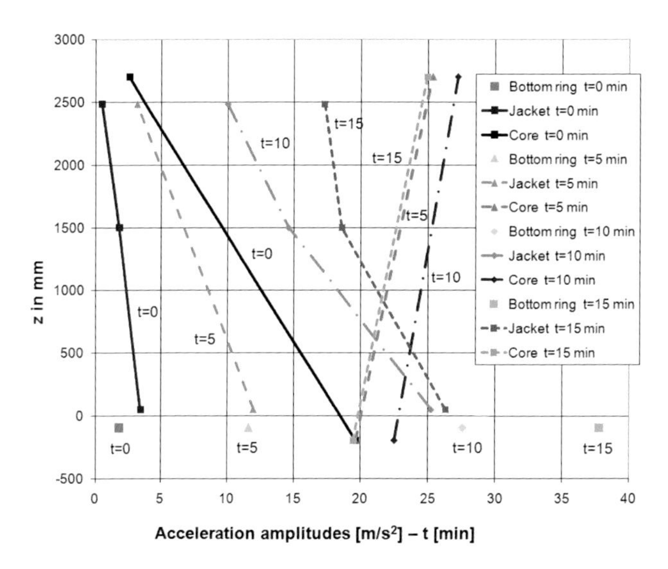

Fig. 4.40:
Vertical distribution of acceleration amplitudes at the core and the jacket depending on the production time of a large pipe

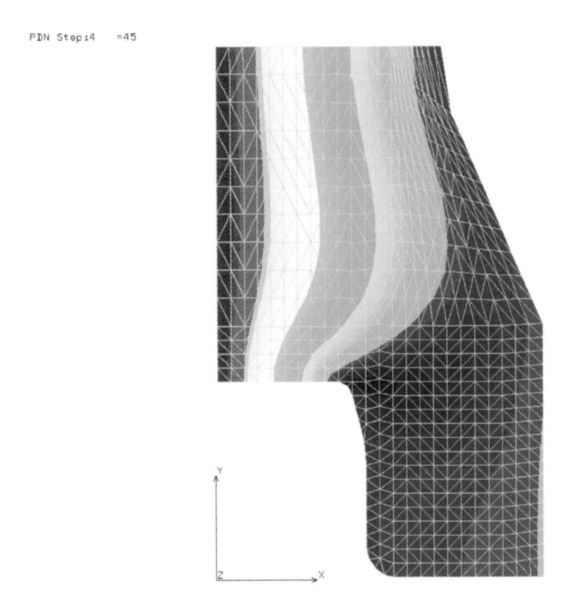

Fig. 4.41:
Distribution of acceleration in the region of the pipe joint; the colours indicate acceleration amplitudes $\hat{a}_x$ [m/s²] in the horizontal direction

Poorly compacted concrete layers that reveal, for example, the accumulation of voids at about two thirds of the pipe height are associated with tilting movements of the jacket and/or the core. The vibration effect is not identical across the individual sections along the pipe. The commonly used settings of the vibration system may result in less significant vibration effects, particularly at about two thirds of the pipe height. Fig. 4.40 shows the results of measurements carried out on a machine producing large pipes. Changes in tilting vibrations during the production process are clearly visible.

A particular challenge is the compaction of the pipe socket, which is precisely where a good compaction quality is necessary to ensure tightness of the pipe connection. Fig. 4.41 shows a cross-section of the vibrated concrete mix in the socket area; the colours indicate the distribution of acceleration. The vibrations introduced from the core do not extend sufficiently into the socket area. A certain portion of the concrete is barely vibrated and remains in place on the bottom ring surface. This is where air voids will be found after the bottom ring has been removed. Low accelerations thus indicate problem areas during compaction of the pipe socket. The distribution of acceleration is governed, and can thus be influenced, by the properties of the concrete mix and the conditions prevailing at the core, bottom ring and jacket, as well as by the excitation frequency.

4.9.1.3 Reinforcement shadows

Integration of the reinforcement is an important step in the production of reinforced concrete pipes. It influences the structural stability and hydraulic functionality of the

Fig. 4.42:
Section of a reinforced concrete pipe with a reinforcement shadow

pipes. One characteristic type of defect is reinforcement shadows below the ring reinforcement (Fig. 4.42).

This phenomenon can be explained by:

- slumping of the concrete due to impacts during transport of freshly demoulded pipes, or concrete settlement as a result of insufficient green strength
- stresses in the reinforcement cage during production, e.g. as a result of the shaping of the spigot end if the concrete cover in the spigot area is insufficient, which can be considered a trivial error
- slumping of poorly compacted concrete during the production process, particularly if vibration equipment is used where deeper concrete layers are also subjected to vibration, which is easily avoided if the concrete layer being cast is sufficiently compacted

Other types of defects whose causes are not yet fully understood occur during incorporation of the reinforcement. They generally result from relative movements between the concrete and the reinforcement or from poor compaction.

The twisting of the reinforcement cage is a further problem. While the jacket and core are vibrating, all points of the jacket and core move on small circular trajectories. This may cause transport processes in the concrete in which the concrete mix slowly rotates around the core. This movement of the mix leads to torsional deformation of the reinforcement cages. After demoulding in the fresh state, the twisted cage makes an attempt to move back into its original position in the fresh concrete, which may also cause reinforcement shadows.

4.9.2 In-Process Quality Control

During moulding and compaction of concrete pipes, it is crucial to exert compacting effects on the concrete mix in order to produce the pipes to the specified quality

Fig. 4.43:
Acceleration sensor mounted on the central vibrator of a pipe machine

standard. In the future, quality control of pipe production will increasingly rely on the consideration of the relevant key parameters. Nowadays, equipment operators are often unaware of the magnitude of these actions in their production processes. These parameters are usually monitored only in exceptional cases.

The compaction equipment thus needs to be fitted with appropriate sensors in order to control key parameters on an ongoing basis. The most basic design of such a system comprises two robust acceleration sensors fitted to the top and bottom ends of the jacket and the core (Fig. 4.43).

If the acceleration parameters are known, the next step is to create a feedback loop to the compaction unit. A control unit checks actual impacts against target values and adjusts the values accordingly. Changes in the conditions during the compaction process can also be taken into account. Such a control system must be supplied with information on system interactions and correlations.

A system to control the compaction unit must also be able to influence the motion parameters of the compaction unit (see also [4.5]).

The properties of the hardened concrete correlate with the bulk densities of the mix achieved during the compaction process. Compaction aims to achieve the specified bulk densities of the concrete. Direct determination of the bulk densities of concrete mixes during the process is not yet possible.

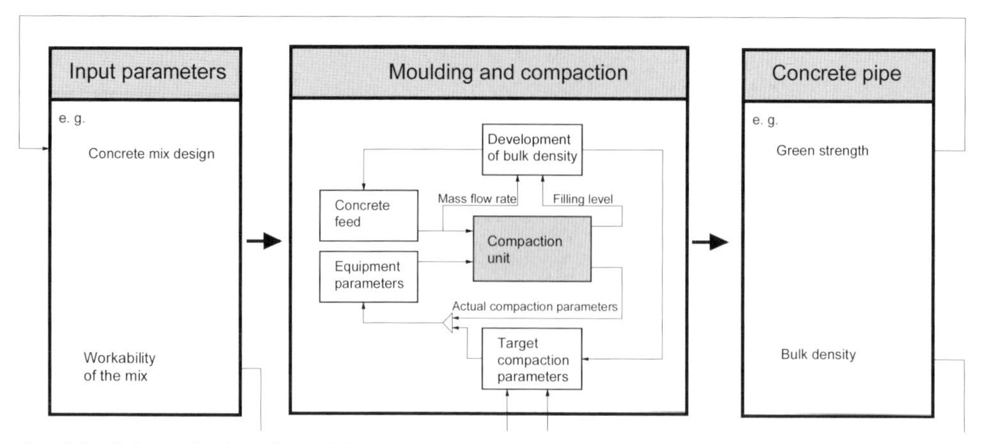

Fig. 4.44: Schematic view of possible automation approaches to quality control of pipe production

Fig. 4.44 shows a schematic view of some of the interactions and correlations relevant to the implementation of automation concepts.

4.10 Selection Criteria

1. Shape and dimensions	Which pipe dimensions are to be manufactured? (nominal bore, pipe length, shape of cross-section, pipe connection) Are there specific requirements as to dimensional tolerances? (jacking pipes) Can the machine produce the required shapes? Is the production process appropriate so that dimensional tolerances can be adhered to consistently? Is the intended location appropriate for the machine? (building height, pit depth, ground conditions)
2. Production output	How many products should be manufactured in which period? (multi-shift operation, depending on dimensions) Can the machine achieve the production output under the prevailing conditions? (in many cases, manufacturer's specifications are based on optimum conditions and the maximum machine availability) Have upstream and downstream processes been designed to achieve this output? (mixer, transport out of the factory, storage facility)
3. Quality	Which parameters are key to the production process? (acceleration amplitude, excitation frequency…) Which parameters are achieved by the machine? Which options exist to influence the production parameters? (frequency converters, unbalance adjustment) Have in-process quality control measures been implemented? Which quality control measures exist for upstream and downstream processes? Are they interlinked?
4. Flexibility	How are moulds exchanged? How much time is required, and what costs does the process incur? Should (and can) various mixes be processed? Are product changeovers and plant extensions possible? (Will the same products still be in demand ten years from now?) Are breakdowns or emergencies possible? (power failure, flood) What happens to the production line in the event of an emergency/breakdown? (emptying of concrete)

5 Production of Precast Elements

5.1 Overview

Precast elements are components ready for assembly that are prefabricated at a factory rather than directly on the construction site. They are transported to the construction site for assembly at a later stage. There are several types of prefabrication, depending on the type of precast elements used [5.3]:

- skeleton construction
- large-panel construction
- mixed construction and special components

In skeleton construction, most of the structural components are bar-shaped (Fig. 5.1), such as columns, posts, girders, purlins, beams, and joists.

Multi-storey buildings also include floor slabs that are either solid for short spans or have a ribbed or hollow-core design for wider spans [5.3]. Skeleton construction is mainly used for industrial and commercial buildings because wide spans are generally required.

Large-panel construction uses precast elements that are mainly panel-shaped (Fig. 5.2). Unlike skeleton construction, the load-bearing vertical elements are wall panels. These wall panels support floor slabs that usually extend over relatively narrow spans.

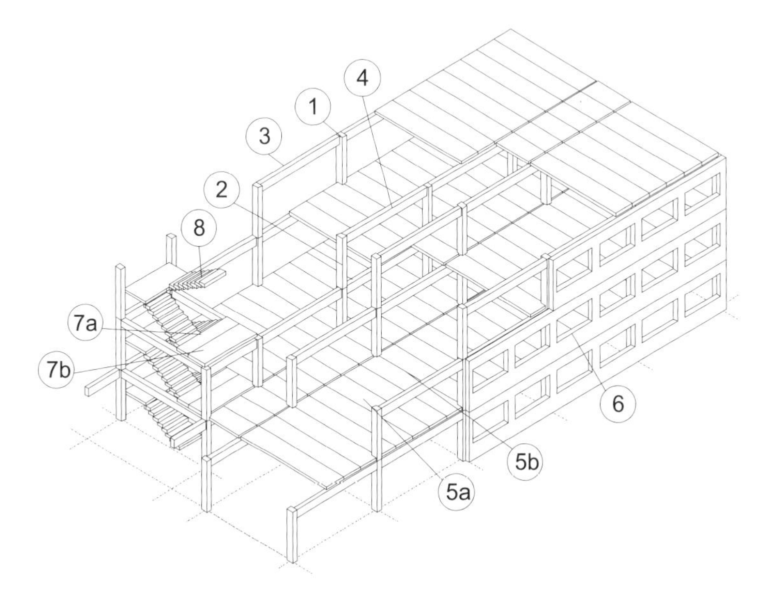

Fig. 5.1:
Typical skeleton structure
1 Perimeter column
2 Internal column
3 Perimeter beam
4 Internal beam
5 Floor slab
(a = bay area,
b = column area)
6 External wall slab
7 Landing slab
(a = slab with support,
b = slab without support)
8 Stair tread

Fig. 5.2:
Typical large-panel building

This construction method is suitable for residential and administrative buildings.

In mixed construction, precast elements are also used for buildings erected in conventional designs. For instance, precast is a cost-efficient alternative for the integration of floor joists, balcony slabs, stairways, landings or other elements into conventional building designs [5.3]. This range may also include special components such as façade panels that are mounted on the building shell using special anchors.

In addition, a wide variety of other special precast elements are used, including bridge girders, tunnel segments, modular units, or box culverts. Reference [5.1] classifies the large number of available precast elements according to product groups (Fig. 5.3).

Table 5.1 shows an example of a concrete mix used for the production of precast elements.

Fig. 5.1: Example of a concrete mix for the production of precast elements

Constituent	Unit	Proportion
Cement CEM I 42.5 R-(ft)	kg/m³	360
Sand 0-2 mm	kg/m³	680
Gravel 2-8 mm	kg/m³	245
Gravel 8-16 mm	kg/m³	839
Additive (AEA)	kg/m³	5.4
Strength class	-	C30/37
Exposure class	-	XC4, XF1, XA1
w/c ratio	-	0.5

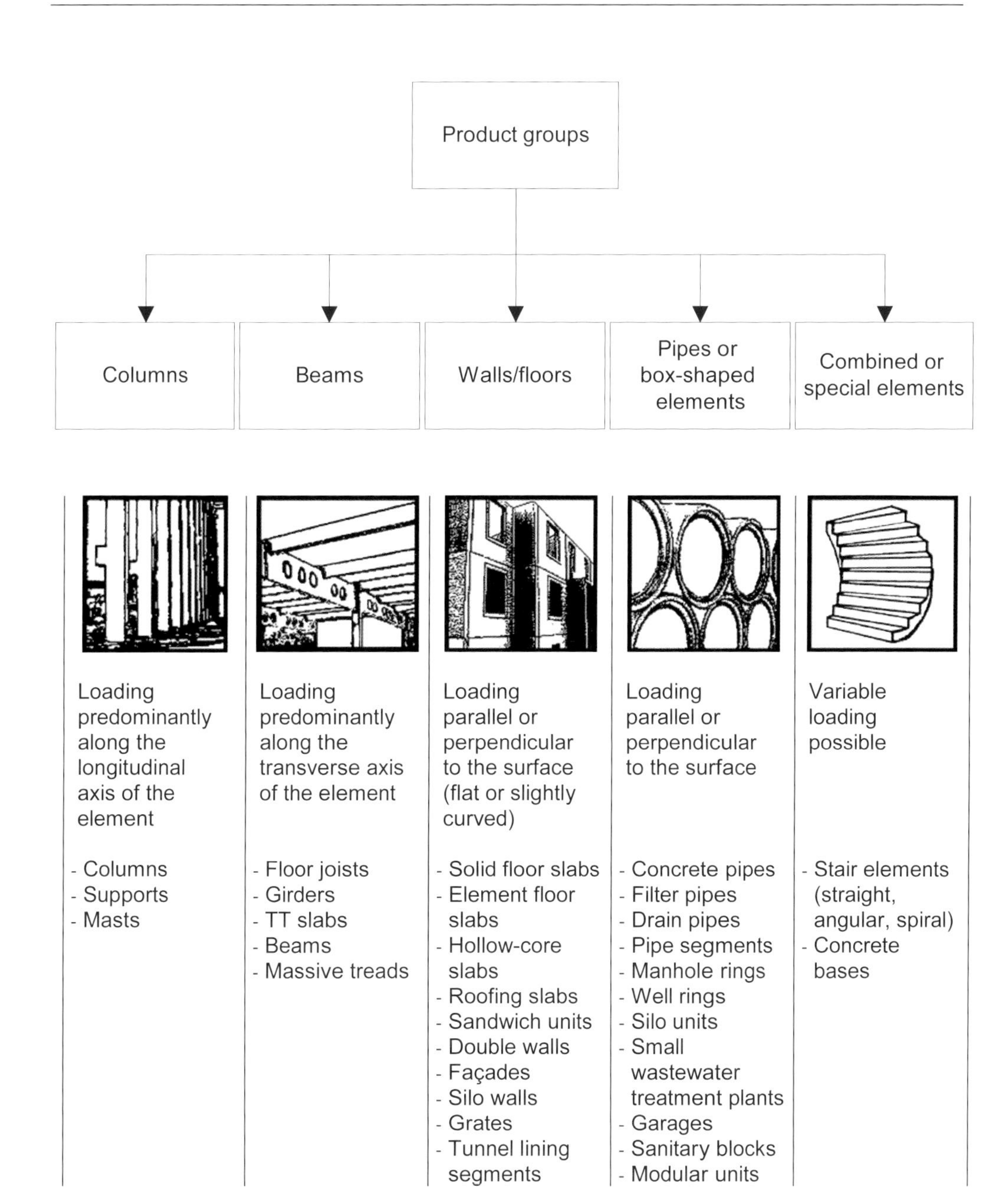

Fig. 5.3: Classification of precast elements according to product groups

5.2 Basic Structure of Production Systems

The production stages of precast elements have already been covered in Section 1.1.3, which also describes the sub-processes involved in manufacturing concrete products.

Section 1.1.2.3 distinguishes between the following two basic organisational principles for production lines:

- carousel
- stationary

Table 5.2 indicates the characteristics of these principles. Various production systems offer technical solutions well-suited to the above-mentioned manufacturing principles.

Table 5.2: Characteristics of production principles

	Carousel production	Stationary production
Work stations	stationary	mobile
Workforce	stationary	mobile
Work equipment	stationary	mobile
Formwork	mobile	stationary

5.3 Carousel Production

5.3.1 Basic Structure

A number of systems implement the carousel principle [5.2]. The following options are available, and a distinction is made according to the direction and layout of the circulation.

a) Direction of circulation:
   - Horizontal circulation (Fig. 5.4)
     Horizontal circulation takes place at a single level.
   - Vertical circulation (Fig. 5.5)
     Vertical circulation involves two levels; the production level is the factory floor and the consolidation level is below floor level. Although this reduces the required floor space by 50%, the construction costs increase, particularly if the groundwater level is high.

b) Layout of circulation
   - Unbranched circulation
     Unbranched circulation involves an essentially rigid coupling of the individual subsystems.

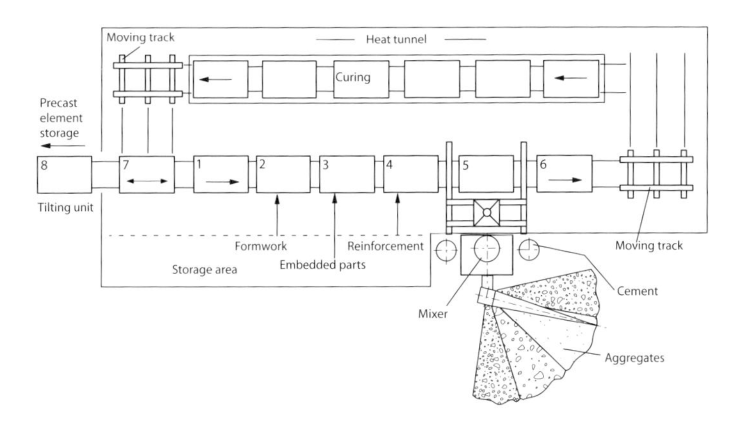

Fig. 5.4:
Horizontal circulation [5.3]

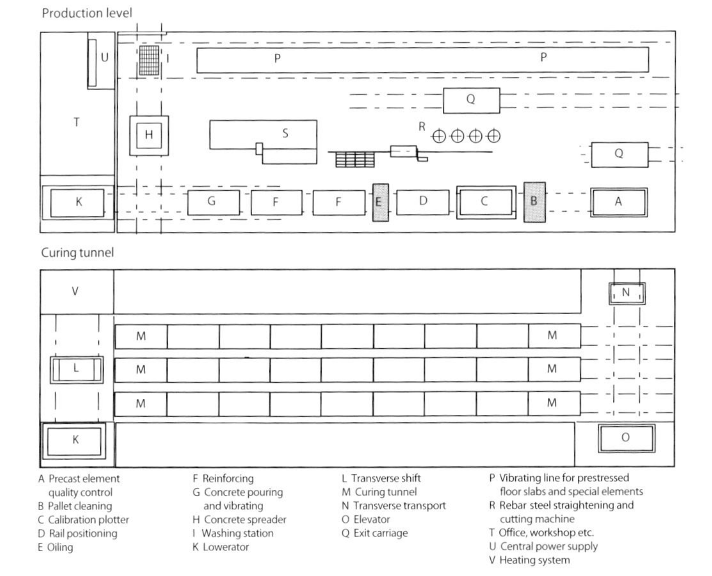

Fig. 5.5:
Vertical circulation

– Branched circulation
  Branched circulation provides a more flexible management and feed of the pallets.

5.3.2 Sub-systems

The main sub-systems of carousel or circulation systems are [5.2]:

– Formwork sub-systems
  - shuttering
  - devices for cleaning and release agent application

  - plotters and/or shuttering robots
  - concrete spreaders
  - compaction units
- Consolidation sub-systems
  - consolidation units
  - deshuttering stations
  - finishing/post-treatment stations
  - exit carriages
- Generally used sub-systems
  - conveyors
  - supply and disposal systems

5.3.2.1 Shuttering

Shuttering pallets are available in two designs with attached side walls:

- shuttering carriages equipped with a friction or chain drive to travel on rails
- shuttering pallets transported from station to station on conveyors.

The design of the side shuttering elements depends on the shape of the product:

a) Formwork for large panels
- side shuttering elements are hinged to the bottom of the formwork (pallet) so that they are foldable
- side elements can be exchanged and are placed onto the pallet and fastened depending on the required height and edge design

Fig. 5.6:
Section of
extrusion formwork

Fig. 5.7:
One method of fixation

- side elements are rigidly connected to the pallet (one longitudinal and one transverse wall); the other sides are flexible so that different lengths and widths can be produced
- side elements are composed of modules and are attached to the pallet.

b) In formwork for parts used in skeleton construction, sections of extrusion formwork are firmly mounted on the pallet (Fig. 5.6).

Because the available space on the pallet is almost never fully utilised due to the shape of the products being manufactured, extra pieces have to be laid between the shuttering elements to act as limiters. Box-outs that require additional shuttering are also often included in the product design. Partitions made from formwork panels are mainly used for this purpose.

The sides and inter-shuttering parts are fastened using the following options:

- detachable mechanical fasteners (Fig. 5.7)
- magnets
- vacuum cups

Magnets and vacuum cups enable formwork to be set-up and removed quickly, accurately and reliably with a low degree of wear.

5.3.2.2 Devices for cleaning and release agent application

After deshuttering, the pieces between the shuttering (transverse shutters, spacers and box-outs) must be removed to prepare the formwork for a new cycle. The shuttering is then cleaned and the release agent applied.

Cleaning machines (Fig. 5.8) are equipped with scrapers and brushes so that the pre-cleaning and final cleaning stages can be merged into a single process.

Fig. 5.8:
Pallet cleaning in a pallet circulation system

Fig. 5.9:
Shuttering robot

An ultra-low volume sprayer is used to apply the release agent. This device creates a thin, uniform film and emits only a very low amount of release agent into the environment.

5.3.2.3 Plotters and shuttering robots

The information system sends the shape and dimensions of the required precast part to the plotter, which then quickly draws the outlines indicating where the shutters are to be placed on the pallet. The shutters can thus be positioned very accurately. If modular shutters are used, their placement can be automated using a shuttering robot (Fig. 5.9). This robot uses sensors to find the individual parts and then takes them out of a magazine.

Fig. 5.10:
Concrete spreader

Fig. 5.11:
Power trowel

5.3.2.4 Concrete spreaders

Pouring and distribution of the concrete within the formwork is carried out with various concrete spreader systems. The conventional method of pouring concrete from a skip suspended from a crane is very flexible. It is still used relatively often.

On the other hand, concrete spreader systems have been developed that often fulfil additional functions. Concrete placement and spreading can thus be supplemented by levelling and smoothing – all these work steps can be carried out by a single system. Concrete spreaders with a portal design are mobile and travel through the factory building on rails or on a frame positioned above the production line (Fig. 5.10). Fig. 5.11 shows a power trowel smoothing the concrete surface.

5.3.2.5 Compaction units

The following technical solutions are used for compacting concrete:

- compaction station with vibrating tables
- mobile machine with surface vibration
- mobile machine with internal vibration
- external vibration on high formwork walls
- internal vibration with manually operated internal vibrators

Vibrating tables are largely independent of the shape of the precast element being produced and can thus be used for a wide variety of purposes. They are the most popular type (Fig. 5.12).

Surface vibration is used only if extensive elements or thin concrete layers are to be compacted because, in these cases, the depth to which the surface vibration effect is achieved is sufficient to compact the product uniformly (Fig. 5.13).

Internal vibrators used to compact the element by hand are very flexible but require a high degree of physical effort to ensure uniform compaction of the entire precast item.

The use of a vibrator cross-beam fitted with several internal vibrators that is lifted into its working position eliminates manual work and thus increases production output.

5.3.2.6 Deshuttering

Precast elements are deshuttered when they have hardened. The side walls are either folded away or pulled off in order to provide enough space to strip the formwork. De-

Fig. 5.12: Shaking table in a pallet circulation system

Fig. 5.13: Surface vibration

Fig. 5.14:
Tilting unit

shuttering stations are usually equipped with stationary pulling devices to remove the side walls.

Wall elements that are not designed to take up loads acting transverse to their surface have to be moved into an almost vertical position before they can be deshuttered. Tilting tables are used for this purpose; the pallets are placed onto these tables and then tilted (Fig. 5.14).

Modular shutters (see Section 5.3.2.1) offer a particularly convenient design because they enable automated manipulation of transverse shutters and their placement in a magazine, as well as their re-positioning on the pallet.

5.3.3 Complete Production Lines using the Carousel Principle

The above-described sub-systems are merged to form complete production lines, which may also combine several circulation systems. The configuration of these lines

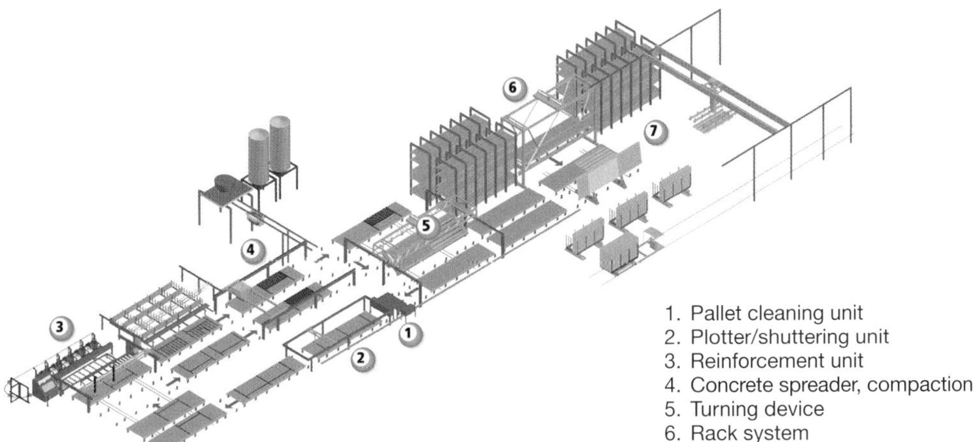

Fig. 5.15: Complete system applying the carousel production principle

depends on the existing factory buildings, the available floor space and their linkage to upstream and downstream processes. Very different layouts are possible using identical sub-systems (Fig. 5.15).

5.4 Stationary Production

5.4.1 Basic Structures

As described in Section 1.1.2.3, the following systems are used to produce large wall panels and elements for skeleton construction:

a) single-mould systems
b) battery mould systems
c) continuous mould systems
d) extrusion systems
e) prestressing line systems (Fig. 5.16)

Each of the overall systems is composed of several (at least two) units: one for moulding and the other for compaction, in alternation.

5.4.2 Sub-systems

5.4.2.1 Single moulds

Single moulds, or single-mould tables, are usually stationary. They can be designed in many different ways both as universal production systems and for specific precast elements. Single moulds are often arranged in series to create entire production lines. For flat-shaped, extensive precast items, two designs are mainly used:

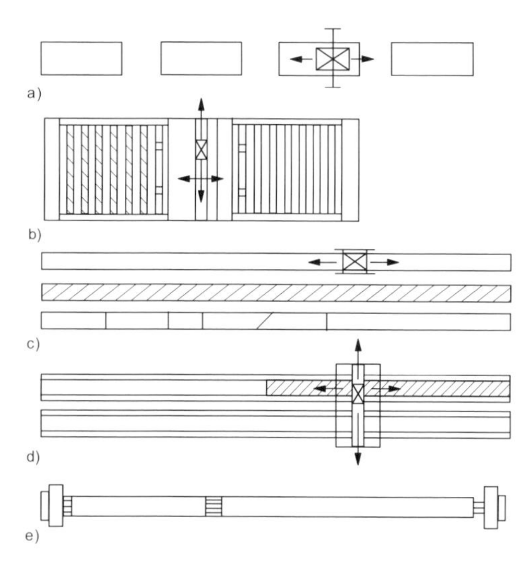

Fig. 5.16:
Stationary production systems
a) single-mould systems
b) battery mould systems
c) continuous mould systems
d) extrusion systems
e) prestressing lines

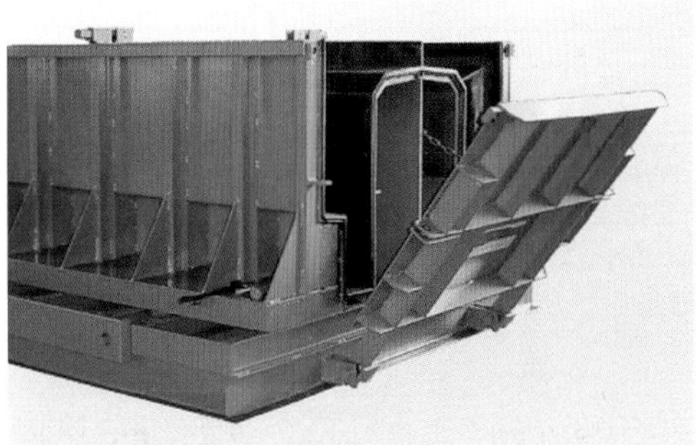

Fig. 5.17: Single mould with a fixed bottom

Fig. 5.18: Tilting mould

- single moulds with fixed bottom (Fig. 5.17)
- tilting moulds (Fig. 5.18)

Single moulds with a fixed bottom are used for all types of flat elements that can be taken directly out of the mould.

Tilting moulds combine a mould and a table. The mould may be made to vibrate independently of the tilt-table frame during the compaction process. Tilting moulds are mainly used for wall units that cannot be demoulded in the horizontal position but have to be moved into an almost vertical position beforehand.

Fig. 5.19:
Mould system for spiral stairs

Fig. 5.20:
Mould for box units

Various single-mould designs are used for special elements and items that are not flat, e.g. stairs, which are key components in both large-panel and skeleton construction. Adjustable stationary stair mould systems produce stairs in various configurations (Fig. 5.19).

Modular items such as box units (Fig. 5.20) or garages are manufactured in vibration mould systems.

Single moulds used for the production of prestressed precast concrete elements include:

- tension-resistant single moulds
- single moulds in a prestressing line

They are suitable for series production of precast elements with low heights and identical dimensions or for prestressed precast elements with relatively short lengths and frequently changing dimensions.

When using tension-resistant single moulds, the mould and the prestressing line are firmly fixed to each other, or abutments act as the faces of the mould. The steel is often prestressed by an electrothermal process outside the mould using heating systems. The elongated prestressed steel strands are then inserted into the prestressing mould and fastened to the transverse abutments. When the concrete has hardened sufficiently, the prestressing steel strands are systematically cut between the mould faces and the transverse abutments in a stepwise process so that the prestressing forces are transferred to the concrete.

Single moulds in short prestressing lines are not subject to stresses. All loads associated with the prestressing process are absorbed by the prestressing line (Fig. 5.21).

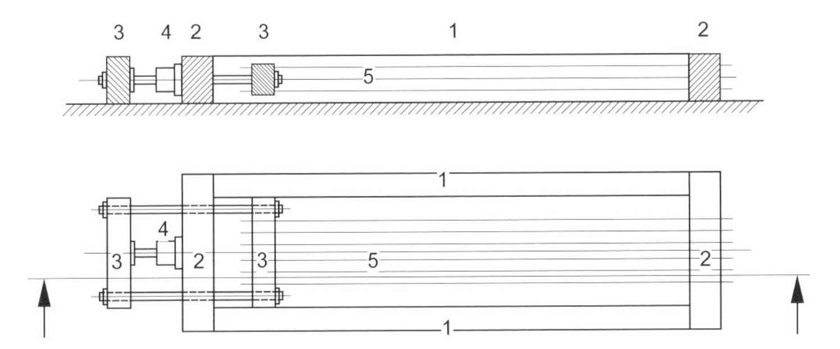

Fig. 5.21:
Short prestressing line
1 Lateral stressing beams
2 Transverse beams
3 Cross-beams
4 Hydraulic press for prestressing the set of strands
5 Prestressing steel strands

The prestressing system is designed as a rigid frame. The prestressing steel strands are anchored to one end and to a cross-beam positioned between the longitudinal beams at the other. The hydraulic press is used to prestress and restructure the steel by the corresponding movements of the cross-beam. Fig. 5.22 shows a short prestressing line for the production of railway sleepers.

Fig. 5.22:
Short prestressing line for the production of railway sleepers

5.4.2.2 Battery moulds

Unlike single moulds, battery moulds can be used to produce several items simultaneously in separate compartments. However, the use of battery moulds is only appropriate for single-layer precast elements, which mainly include interior and basement wall units. This method of producing precast elements offers a number of advantages:

- production in the position for subsequent transport and final installation
- smooth, even surfaces by bilateral mould enclosure
- lower costs of all work steps
- favourable conditions for heat treatment provided by the compact coupling of the individual mould compartments to a battery (low heat loss)
- small footprint [5.2]

Battery moulds are available in various designs; however, they are generally composed of a series of identical assemblies arranged in parallel. The compartments for the pre-

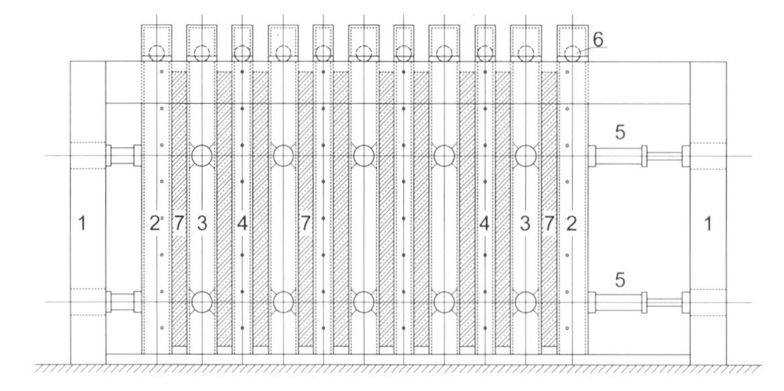

Fig. 5.23:
Battery mould
with top suspension
1 Support frame
2 End walls
3 Partition wall as a vibrating element
4 Partition wall as a heating element
5 Rams
6 Roller bearing
7 Concrete precasts

cast element are located between partition walls. The end walls are fitted with supports that hold the entire structure together. Mould bottom and sides are positioned between the partition walls. Two different systems are commonly used to tilt the partition walls into a vertical position and to transport them:

- top-suspended using rollers and a support frame
- base-mounted using rail-guided rollers

Precast elements must be lifted upwards out of top-suspended battery moulds (Fig. 5.23), which requires a sufficiently long crane hook. Another option is to position the battery mould as a single group below floor level. Both solutions require a considerable construction outlay.

Precast elements are removed from base-mounted battery moulds from the side with only a little lifting because this design does not have a longitudinal frame. Standard crane hooks are usually sufficient. The tie rods holding the battery mould firmly together can be released prior to demoulding (Fig. 5.24).

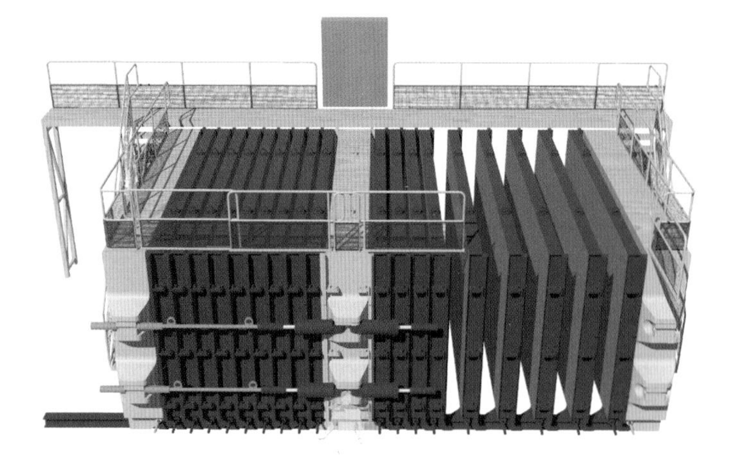

Fig. 5.24:
Base-mounted
battery mould

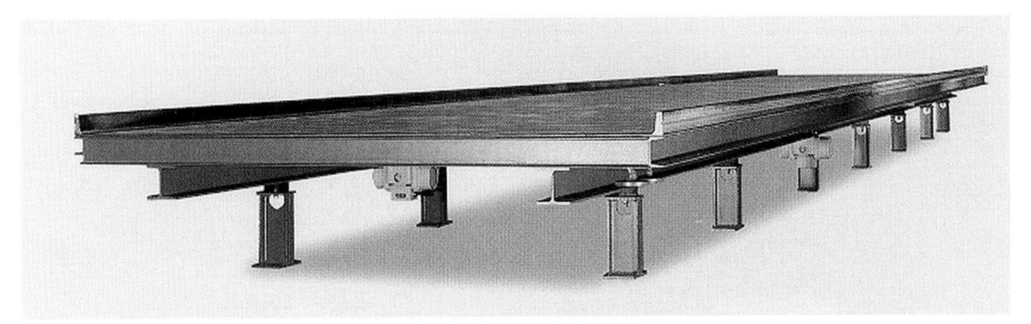

Fig. 5.25: Continuous mould

A mobile concrete hopper travelling above the mould pours the concrete into the mould compartments. The concrete is then compacted by vibrators integrated into every second partition or by external vibrators mounted on the end walls.

Every other partition wall accommodates a pipe system that distributes steam or thermal oil for heating purposes. The heat treatment process is thus based on contact with the concrete via the surfaces of the heated partition walls.

In the demoulding process, the restraints are loosened and the partition and end walls are removed. The precast elements can then be taken out and placed in racks close to the battery mould for intermediate storage and post-treatment as well as for inspection and acclimatisation purposes.

5.4.2.3 Continuous moulds

Continuous moulds are stationary mould tables that are fitted with a continuous mould bottom that extends over a great length (Fig. 5.25). They are used to produce large, flat precast items in varying lengths and finishes, such as lattice-girder and hollow-core floor slabs.

The length of the individual precast items is limited by transverse shutters arranged perpendicularly or at a different angle to the longitudinal axis of the continuous mould. This provides flexibility in terms of the desired element layout or shape.

The concrete is poured through concrete hoppers travelling above the continuous mould. In most cases, the concrete is compacted by external vibrators fitted underneath the mould. The problems caused by this arrangement are discussed in detail in Section 5.7.

Elements with level surfaces can be produced by machines that combine a concrete hopper and surface vibrator in a single unit.

Fig. 5.26:
Girder mould

5.4.2.4 Extrusion moulds

Extrusion systems are moulds used for the production of precast elements for skeleton construction. Similar to continuous moulds, they extend over a great length and are thus well-suited to the production of beams, columns and girders, which cannot be manufactured in a carousel or circulation system due to their heavy weight.

These extrusion moulds are mainly available in three versions:

- mould for beams and columns
- moulds for girders (Fig. 5.26)
- mould for TT slabs (Fig. 5.27)

The height and width of these moulds are adjustable so that they can be used flexibly for various cross-sectional dimensions and shapes. The mould design includes a base frame, with side walls and end walls positioned on top of it. The formwork facing is generally made of wear-resistant sheet steel. Wooden formwork panels are only used for special modifications.

Fig. 5.27:
Mould for TT slabs

Concrete is poured and spread from a concrete hopper or crane skip. External vibrators mounted on the side walls are generally used for compaction.

The concrete is usually cured by pipe systems installed underneath the mould bottom and filled with a heat transfer medium.

The adjustable side walls of the mould provide sufficient space for demoulding.

5.4.2.5 Prestressing lines

Prestressing lines or beds enable the production of several consecutive prestressed concrete items with an immediate bond. These lines can be more than 100 metres long, and in contrast to the short prestressing lines described in Section 5.4.2.1, their abutments have deep foundations. Prestressing cross-beams enable tensioning of several strands at the same time (Fig. 5.28).

A typical example of the extrusion on prestressing lines is the manufacture of hollow-core floor slabs. Hollow-core slabs are long structural members with longitudinal reinforcement and moulded cavities. The production processes shown in Fig. 5.29 are generally appropriate for this purpose.

The particular advantages of this production system include a consistently high quality, reliable scheduling and cost effectiveness, which are also general advantages of precast production. Prestressing allows a relatively low structural height. Due to the cavities incorporated in the element, 50% less material is consumed compared to an equivalent floor cast in-situ.

In a typical hollow-core cross-section, the reinforcing steel required for structural reasons is inserted in the bottom layer.

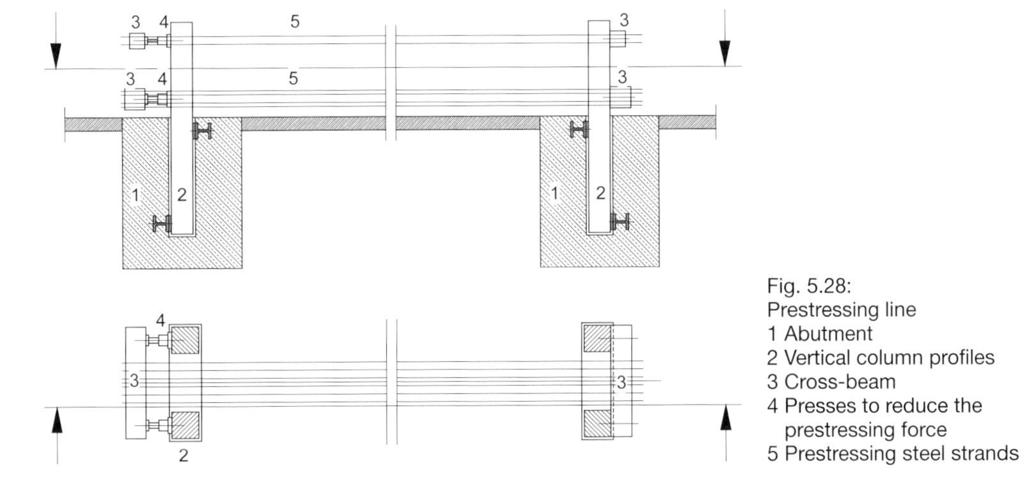

Fig. 5.28:
Prestressing line
1 Abutment
2 Vertical column profiles
3 Cross-beam
4 Presses to reduce the prestressing force
5 Prestressing steel strands

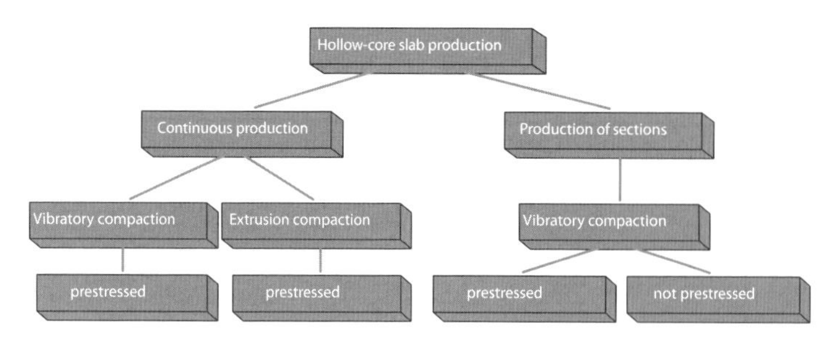

Fig. 5.29:
Production process for hollow-core slabs depending on precast element length, compaction method and insertion of reinforcement

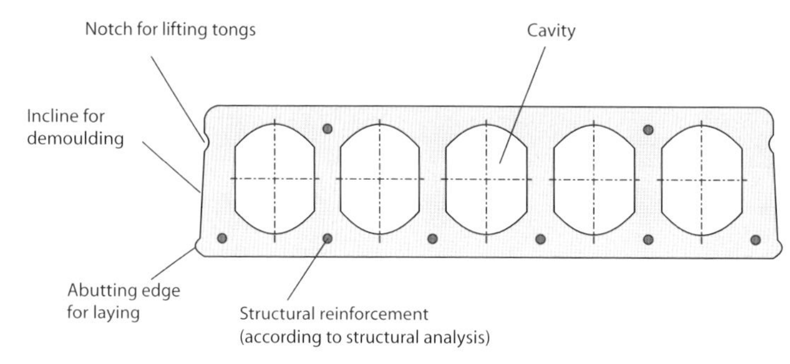

Fig. 5.30:
Functional details of a hollow-core slab design

The cavities are located above the reinforcement. Depending on the floor thickness, additional strands may also be inserted between the cavities. They increase the inherent stability of the element, primarily during its transport within the factory or to the construction site. Fig. 5.30 shows a cross-section indicating the functional components.

Specially designed floor units are fitted with a thermal insulation layer. In this case, foamed polystyrene parts are inserted into the prestressing bed and covered with concrete (Fig. 5.31). This avoids placement of insulation at the construction site.

Two moulding and compaction processes are widely used in practice: slipforming and extrusion. In the slipforming process, the concrete mix is conveyed into the mould

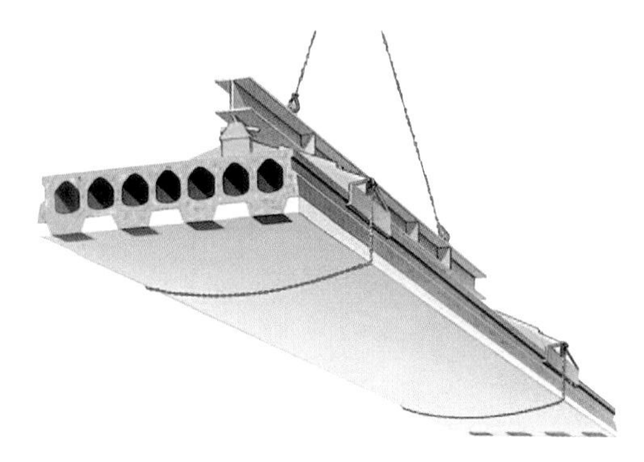

Fig. 5.31:
Design of an insulated hollow-core slab

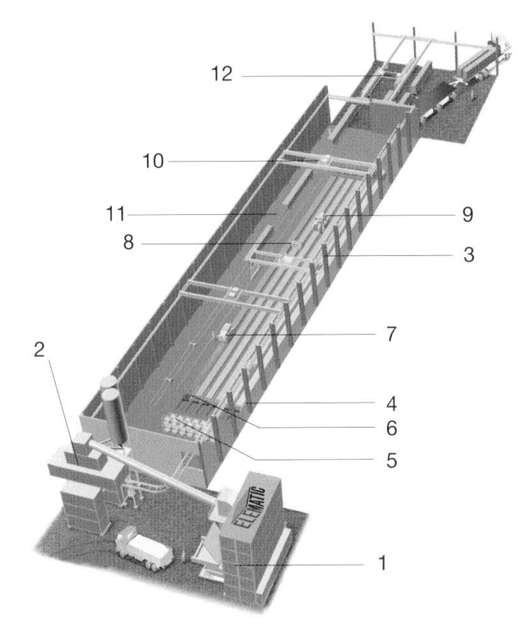

Fig. 5.32:
Plant and equipment for the extrusion process
1 Aggregate storage
2 Mixer
3 Concrete spreader
4 Prestressing bed
5 Strand storage
6 Prestressing
7 Extruder
8 Plotter
9 Saw
10 Crane
11 Element transport
12 Storage

opening of the slipformer. A series of cores are used to form cavities in the slab. The concrete is compacted by vibration of the cores or of tools positioned on top. The slipformer is moved by its own drive.

Even stiffer mixes can usually be processed by extrusion. In this process, the concrete mix is conveyed into a chamber with conical screws within the production machine. The screws rotate to transport the mix towards the greater screw diameter and to compact it as a result of the decreasing cross-section. The compacted mix is pressed into the remaining space formed by the side and top mould walls because the cores are located immediately adjacent to the screw ends. This extrusion pressure provides the required advance rate of the machine.

Extrusion equipment for prestressing lines
Production takes place in buildings that are more than 100 metres long. Several lines are usually arranged next to each other. While elements are being produced on one half of the lines, the other half is used to remove the elements and to prepare a new production cycle. Fig. 5.32 shows the plant and equipment required for the extrusion process.

Slipformers
Slipformers include the following main components: concrete silo, compaction unit, moulding unit, frame with drives and controller (Fig. 5.33).

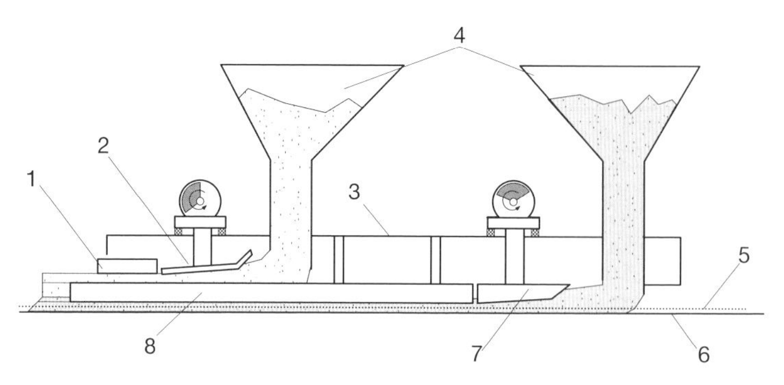

Fig. 5.33:
Schematic view of a slipformer
1 Levelling screed
2 Vibration screed
3 Frame, drives and undercarriage
4 Concrete silo
5 Steel strand
6 Bottom of prestressing line
7 Vibration screed with cavity contour
8 Core

The concrete silo may be divided into several chambers to enable gradual compaction. The thickness of the layer worked by the tool is thus limited, which prevents problems that may otherwise arise because the vibration effect does not reach the full depth of the element.

Compaction is based on vibration. Various solutions are used in practice. For instance, the concrete mix is compacted by top-mounted tamper heads or vibration screeds. If the concrete is compacted in several layers, the structural reinforcement is usually inserted in the first layer. Depending on the thickness of the slab, up to three concrete layers may have to be compacted individually. Vibration may also be used for the cores (Fig. 5.34).

A levelling screed is positioned downstream of the compaction step to smoothen the surface. This screed may either be passive or oscillate horizontally transverse to the direction of travel. Depending on the specifications, the surface can be roughened again to enable a more effective bond to the portion of the concrete that will be cast at the construction site.

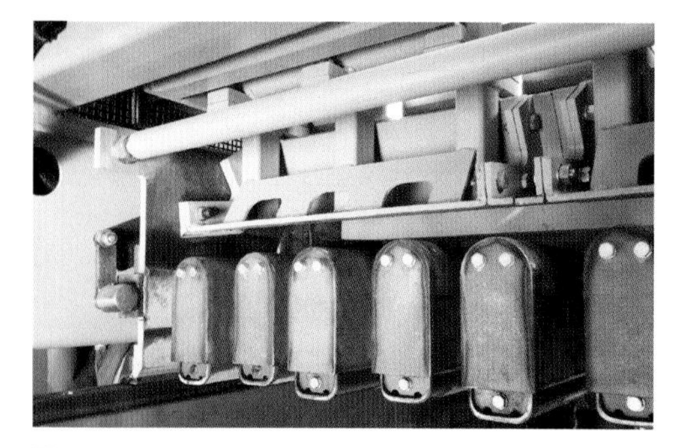

Fig. 5.34:
Cores to shape the cavities in a slipformer; a levelling unit is positioned on top

Fig. 5.35:
Slipformer

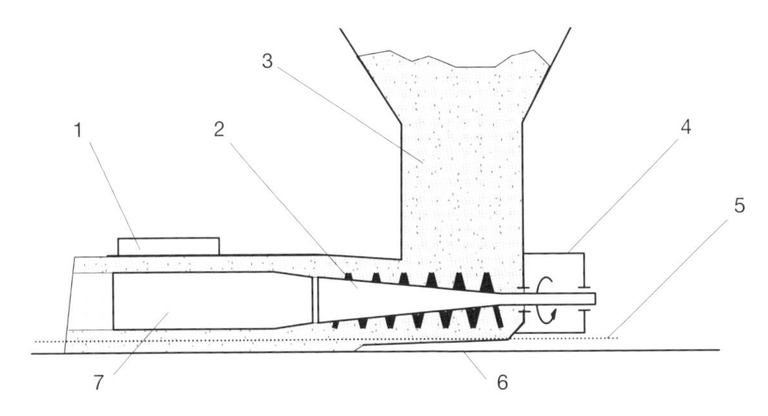

Fig. 5.36:
Schematic view of an extruder
1 Levelling screed
2 Extruder screw
3 Concrete silo
4 Frame, drives and undercarriage
5 Steel strand
6 Bottom of prestressing line
7 Core

Slipformers are commercially available with a standard width of 1.20 metres and also in working widths of up to 2.40 m. The latter may also be used to manufacture precast items that are divided in the longitudinal direction.

Slipformers usually advance at a speed of 1 to 3 m/min.

Extruders

Extruders use the principle of shear compaction. The mobile compaction unit has a design similar to that of the slipformer. The frame with the undercarriage supports a concrete silo from which the mix is conveyed into the compaction unit (extruder), where it is pressed into the mould by a reduction in the cross-section and friction. Fig. 5.36 illustrates the working principle of an extruder.

The final moulding of the cavities is performed by cores that are positioned immediately downstream of the extruder screws (Fig. 5.37). Other tool shapes or movements

Fig. 5.37:
Extruder

may be included for compaction, e.g. in the longitudinal direction. Extruders may also use core vibration or vibration screeds to assist compaction.

The extruded mix is very stiff because of the process design. Concrete with a w/c ratio between 0.32 and 0.36 is poured. Nowadays, wear of the extruder unit is reduced by using special materials for the screws. According to manufacturers' statements, over 60,000 m^2 of hollow-core slabs can be produced without having to replace any wear part.

Peripheral equipment
Apart from the prestressing bed, several pieces of peripheral equipment are required. These include:

- prestressing device
- unit to unwind the reinforcing steel
- plotter
- saws
- equipment to clean and prepare the prestressing line
- concrete spreader

5.4.3 Complete Lines for Stationary Production

The individual sub-systems for stationary production are combined in complete systems in which each sub-system remains relatively independent. This principle enables a high degree of flexibility in stationary production lines. Several sub-systems are usually arranged in series and are merged in a single section of the production line. This

Fig. 5.38:
Single-mould production line

enables separation of the moulding and compaction steps referred to above, which is beneficial given the differing duration of these sub-processes.

A serial arrangement is thus appropriate for both single moulds of any type and battery moulds. One compact serial arrangement of battery moulds is the so-called dual mould, where two battery moulds are coupled to each other to form a processing unit. They use the same concrete spreader and are divided into moulding and compaction in an alternating pattern (Fig. 5.38).

Several continuous moulds are positioned parallel to each other at the working distance to form an overall continuous system.

In the same way, complete extrusion systems are created by arranging several extrusion moulds parallel to each other.

The possible number of continuous moulds or extrusion lines that can be positioned next to each other depends on the width or span of the factory building. The number of required moulds is derived from the required production output.

5.5 Combined Production

Precast plants almost always include several production lines. In most cases, the following lines are combined:

- production of large panels and skeleton units
- carousel and stationary production

This is due to the wide variety of products that is often included in the order, which results from the structural layout of the specified building and includes both large panels and skeleton elements. Requested items thus include walls, floors, columns and beams, as well as girders in commercial or industrial buildings. These requirements lead to a division of the factory into clearly delineated production areas that are tailored to the manufacture of certain types of precast elements.

Depending on the size and type of the precast elements, a combination of carousel and stationary production may also be implemented where, for instance, shorter columns or beams up to the length of the mould carriage are produced in a circulation system whereas longer elements are manufactured in a stationary arrangement.

Precast plants with a focus on carousel production usually also include some stationary production facilities to supply ordered products that cannot be manufactured in a circulation system. Plants that use mainly stationary equipment and additional carousel production systems are less common than the other way round.

Systems using the individual production principles may also be combined. Especially in stationary production, single and continuous mould systems and extrusion lines are often installed in a single precast plant.

5.6 Curing and Finishing

5.6.1 Curing Systems

Curing is a time-consuming process, which is why the concrete-filled moulds remain in the production system for a comparatively long period. This process step requires relatively large dedicated areas.

Curing systems include [5.2]:

- curing tunnels (Fig. 5.39a)
- curing racks (Fig. 5.39b)
- designated curing areas (Fig. 5.39c)

Curing tunnels enable an essentially continuous flow. Moulds are fed into the tunnel in sync with the production cycle. They pass through the tunnel and are discharged in the next cycle. Elevators and lowerators are required for multi-level tunnels.

Curing racks are multi-level systems in which the moulds remain stationary during the curing period. These installations also require elevators and lowerators. Fig. 5.40 shows a heated curing chamber with an automatic storage and retrieval unit that is incorporated in a circulation system.

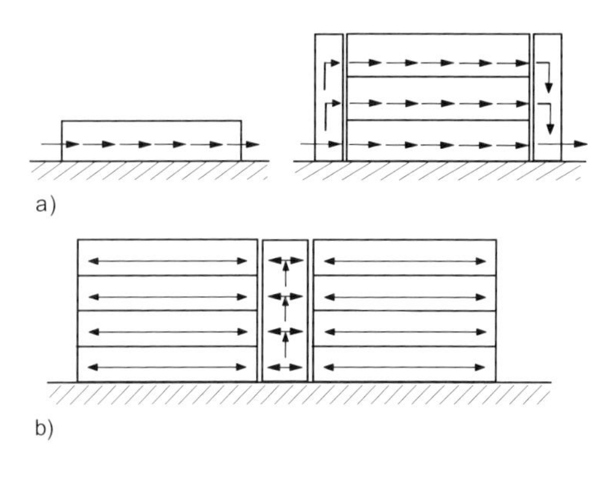

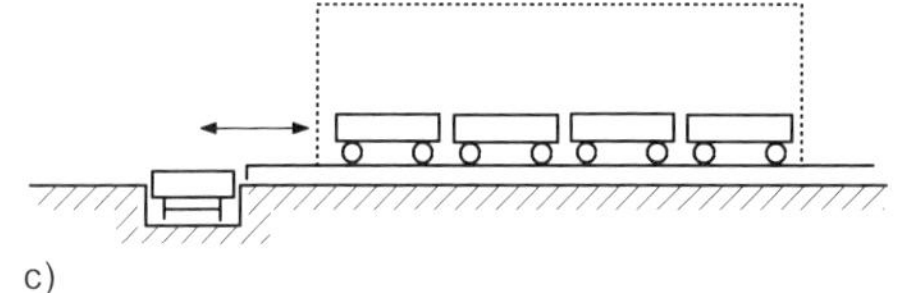

Fig. 5.39: Curing systems
a) Curing tunnel
b) Curing racks
c) Designated curing areas

Fig. 5.40: Heated curing chamber with automatic storage and retrieval unit

A designated curing area is the simplest but most space-consuming option. The moulds are placed on an area that is often simply covered with tarpaulins.

In the above systems, the curing process is usually accelerated by heating, or the heat generated during the hydration process is used for this purpose. Possible heat-transfer mediums include:

- steam in heat-transfer pipe systems
- mixtures of wet steam and air for direct treatment
- hot air

5.6.2 Finishing

Finishing is mainly used for façade panels (external wall elements). Processes include

- washing
- grinding

Washing stations include a washing unit and a frame that holds the elements in an almost vertical, slightly inclined position. Through a controlled movement of the spray nozzle, the washing unit applies water to the concrete surface whose curing was delayed deliberately. In this process, the cement is washed out to a certain depth (Fig. 5.41).

In some cases, this process is carried out manually and assisted by brushing. The water used for the washing process is recovered, cleaned and recycled.

Grinding stations for finishing of façade panels consist of grinding machines with a portal that moves in the longitudinal direction. The grinding tool is suspended from this portal so that it can move in the transverse and vertical directions [5.2]. There are

Fig. 5.41: Washing station

Fig. 5.42: Grinding machine

grinding machines for surfaces and edges. Using different grinding heads, a grinding machine designed for surfaces can also be used for rough and fine grinding of (oblique) edges (Fig. 5.42).

Edge grinders are suitable for finishing window and door reveals, shadow gaps and offsets. They are equipped with relatively small heads and exchangeable tools.

Utmost care must be taken during mould preparation, concrete pouring and demoulding to achieve a good finishing result.

The water used for grinding is collected in tanks, cleaned and recycled.

5.7 Quality Control

In addition to the composition of the concrete mix, its high-quality processing and uniform filling of the moulds, the moulding and compaction stage is one of the key subprocesses in the production of precast elements (see Section 1.1.3).

As described in Section 1.1.4.2, it is crucial that the vibration parameters of the compaction unit, which were previously matched to the concrete mix both theoretically and in tests, are introduced uniformly into the mix at all points of contact between the mould and the concrete. Furthermore, care must be taken to ensure that the vibration energy is transferred further into the mix so that a uniformly high quality is created across the entire cross-section of the precast element. Despite the development and application of self-compacting concrete (SCC) and easily compactable concrete (ECC), vibration continues to be the method that is most frequently used in precast element production.

5.7.1 Design of Vibration Moulds

Mould systems required for moulding and compaction of precast elements are composed of large steel structures that have to be excited in order to vibrate. This vibration energy is then introduced into the mix at appropriate contact areas. For example, these systems are used to produce flat precast elements such as walls and floor slabs using tilt tables, battery moulds and vibrating units and for moulding and compacting curved elements such as tunnel lining segments, pipes and garages, as well as for the manufacture of large, complex components for industrial buildings.

5.7.1.1 Systematic classification of vibration moulds

There is no consistent and comprehensive systematic classification of the above-described vibration mould systems. Grouping according to the types of concrete products or product groups (Fig. 5.3) appears feasible to a certain extent. For example, this system is used to make a distinction between moulds for tunnel lining segments, pipes, garages, box units and stairs. For the purpose of the subsequent model identification and analysis, a structural classification is assumed (Fig. 5.43).

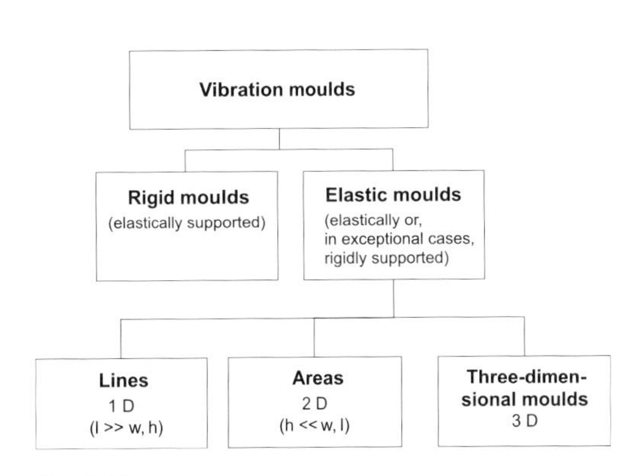

Fig. 5.43:
Classification of vibration moulds

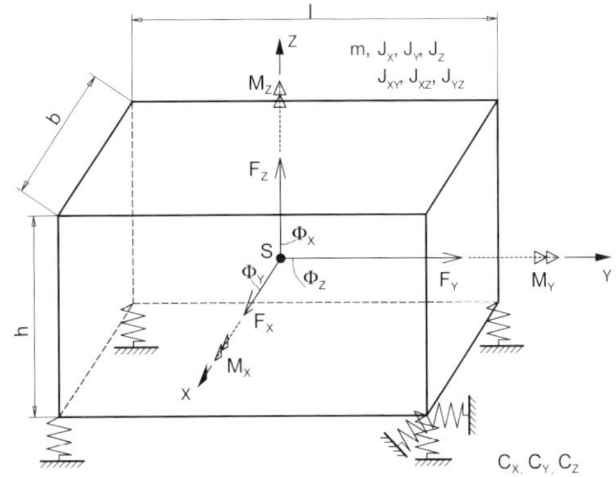

Fig. 5.44:
Rigid-body model and main dimensions of a simple box mould

The physical parameters given in Fig. 5.44 are used. The categories of tilt tables, pallets, battery moulds or continuous moulds are then discussed. Elastic moulds account for the largest share in the above-mentioned vibration mould systems.

5.7.1.2 Dynamic modelling and simulation

A safe approach to meeting the above-mentioned requirements with respect to the introduction of vibration energy into the concrete mix is dynamic modelling and simulation of the equipment (see also Section 1.4.2). The software packages used for this purpose are the Adams multi-body simulation (MBS) suite and the finite-element method (FEM) software Cosmos M and Cosmos Design Star.

Like all dynamic systems, vibration moulds are vibrating systems whose mathematical models include mass properties, elastic characteristics, damping properties and the vibrational excitation [5.5].

For elastically supported rigid moulds, it is relatively simple to create the model and carry out the analysis because the known calculation model for an elastically supported rigid body with six degrees of freedom can be applied (Fig. 5.45). A mould is considered rigid when its natural deformation frequencies are much greater than the excitation frequencies.

Even this simple rigid-body model can identify a significant number of parameters that influence the motion characteristics of the rigid mould. These include the geometrical dimensions of the mould, the masses and/or their distributions, the point of

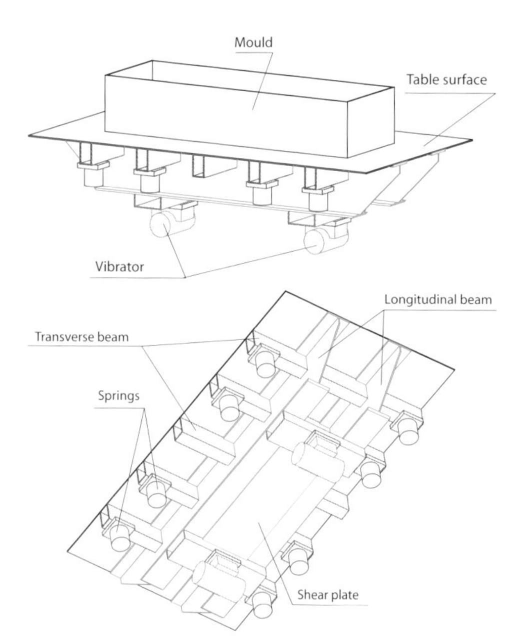

Fig. 5.45:
Vibrating table: general configuration (structure, main components)

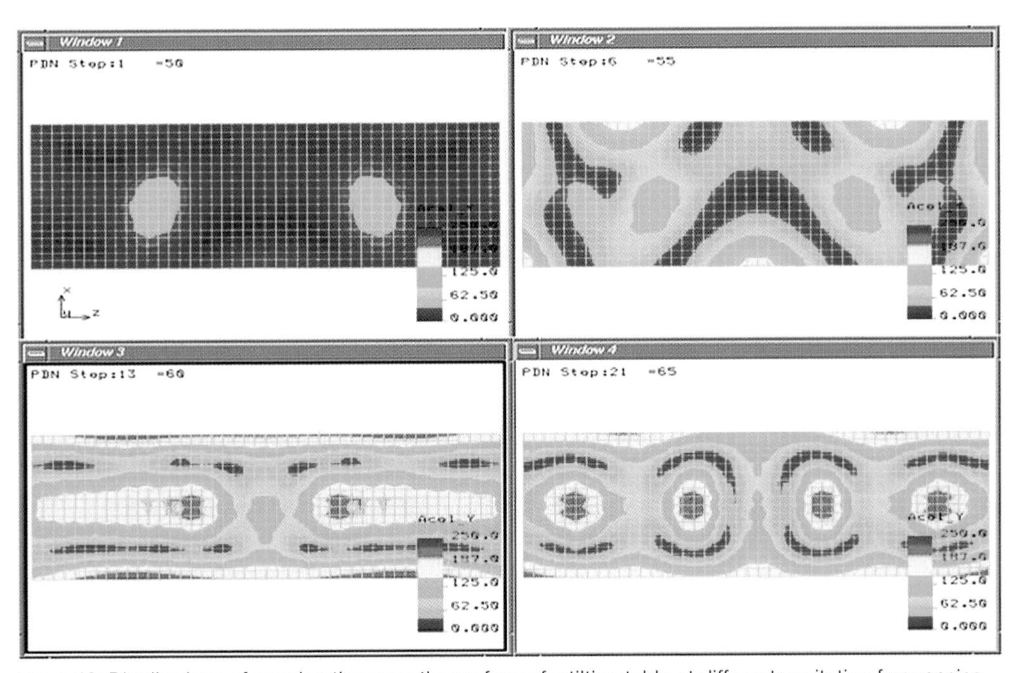

Fig. 5.46: Distributions of acceleration over the surface of a tilting table at different excitation frequencies

introduction and frequency of excitation as well as the elastic support with its points of contact.

However, vibration moulds are essentially elastic structures, which is why the model must reflect the vibration mould in its design as an elastic system capable of inherent vibration. Modelling and analysis using the finite-element method is recommended for such cases. In the course of this procedure, the structure is transformed into an abstract model having a sufficient degree of accuracy and broken down to small sections (finite elements). The corresponding material characteristics are then allocated to these elements. This mathematical model can be used to calculate the vibration parameters relevant to vibratory compaction and their distribution at the points of introduction into the concrete mix. For example, Fig. 5.46 shows the distribution of acceleration over the surface of a tilt table during excitation by several external vibrators.

The clearly visible non-uniform distribution of acceleration over the surface inevitably results in corresponding differences in density, and thus leads to concrete compressive strengths that vary across the surface of the precast element. However, this example also shows that any attempt to use a rigid structure to trigger a co-phasal, homogeneous vibrational movement of the entire area is bound to be unsuccessful in the case of certain geometrical dimensions of the vibration mould system in conjunction with its excitation frequency. Hence, the important aspect is to use a model calculation to find a parameter distribution that is uniform across the entire area. A corresponding parameter that represents such a sufficiently uniform nature of the vibration parameter has been defined in [5.6].

Of course there are also special cases involving complex vibration moulds in which it might be advantageous to design at least some sub-systems as rigid structures. In this case, too, dynamic modelling of the equipment is a very useful aid in decision-making.

Model identification and analysis considers all key parameters that influence the motion characteristics of the vibration mould, therefore these influences can also be analysed in terms of their effects, and matched accordingly. Examples include structural details such as sheet thicknesses, profile cross-sections or grid dimensions, as well as the type (spring stiffener), number and position of spring isolators, and size (excitation force), frequency range and location of vibrators. The enormous influence of the excitation frequency on the distribution of acceleration over the surface of a tilt table is clearly visible in Fig. 5.46.

The above-described options for model identification and analysis make it possible to simulate the motion behaviour of conventional vibration moulds that have been tried and tested in practice, and to calculate the loads, stresses and deformations acting or occurring in this process. What is more important, however, is to utilise the existing simulation options to develop novel technical solutions for moulding and compaction of concrete mixes.

5.7.1.3 Innovative technical solutions

In addition to improving existing equipment by performing the studies referred to above, a number of completely new technical solutions have also been developed to meet the aforementioned requirements with respect to moulding and compaction of concrete mixes. These relate to the uniform introduction of vibration energy at all points of contact with the concrete mix, the uniform transfer of vibration energy over the entire cross-section of the concrete product, and the flexibility of the moulding and compaction equipment. New solutions were also developed to mitigate unwanted noise and vibration occurring during moulding and compaction. The following section describes an example of such a new solution for the uniform introduction of vibration energy into the concrete mix [5.4].

As mentioned above, it is necessary to introduce vibration energy uniformly at all areas of contact with the concrete mix in order to achieve a high product quality. As already outlined, this is not possible with the commonly used flexible vibration equipment. The distribution of vibration is determined not only by the table design, but also by the excitation frequency, element shape and concrete mix composition.

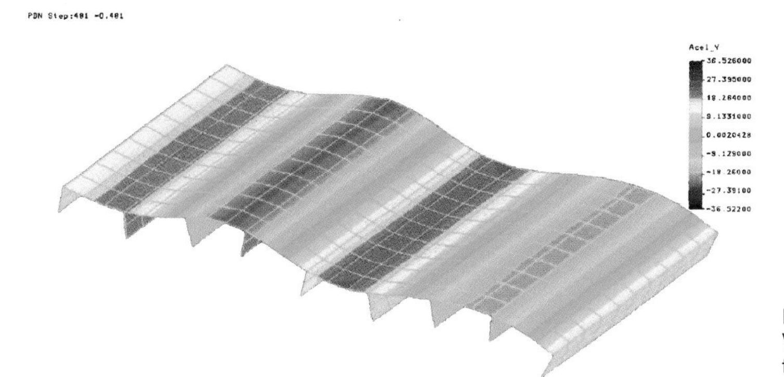

Fig. 5.47:
Vibration mode on the undulating table

This problem has been addressed by developing the innovative undulating table design. The exciter system transmits a transverse wave to the vibration mould. This results in acceleration amplitudes that are uniform across the entire surface of the vibration mould, with the acceleration peaks measured at different points in time. Fig. 5.47 shows the wave-like motion of the table surface.

The table consists of an elastically supported, flexible mould area with longitudinal beams. Two drive trains with unbalances in a phase-shifted arrangement are mounted on the sides (Fig. 5.48).

Based on this functional principle, a modular system with standard components can be used to design a customised undulating table. Fig. 5.49 shows an undulating table with a mould surface of two by four metres.

This table not only achieves a high product quality, but it also ensures low-noise operation. Reference [5.4] describes other innovations for manufacturing precast elements:

Fig. 5.48: Undulating table with two drive trains

Fig. 5.49: Undulating table in the concrete plant

- use of spherical vibration for three-dimensional introduction of vibration energy into the concrete mix for precast elements with a complex shape
- the flexible, modular Flexmodul moulding and compaction system, which consists of:
  - vibrating table
  - suspension system
  - excitation system
  - controller

and:
- low-frequency, three-dimensional introduction of vibration energy to achieve
  - a high surface quality
  - flexibility with respect to the dimensions of the element
  - noise reduction

5.7.2 In-Process Quality Control

The principles of quality control are described in Section 3.2.4. This section evaluates the question as to whether the requirements of uniform introduction of vibration from the mould to the concrete mix and its uniform transfer within the mix can actually be met by the vibration equipment used. In many cases, problems arise during moulding and compaction of precast elements that result in related quality defects, which is why measurements need to be carried out on vibration mould systems to capture their motion behaviour.

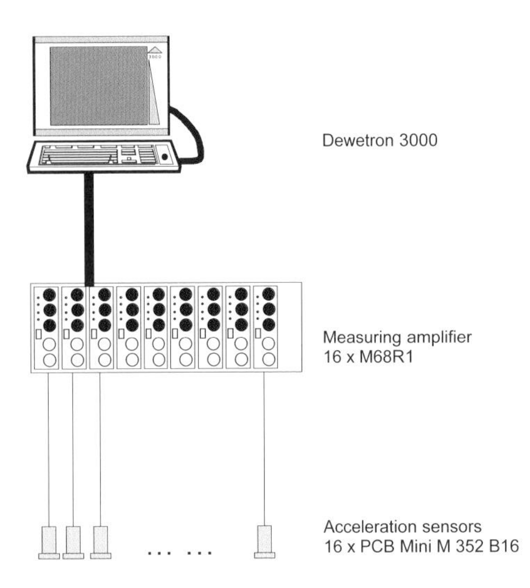

Fig. 5.50:
Device configuration to capture and analyse measured values

Fig. 5.51:
Vibration measurement technology in use

Fig. 5.52:
Arrangement of acceleration sensors on the production pallet of a circulation system

Such measurements also verify the appropriate isolation of the vibratory compaction system from its surroundings. Fig. 5.50 illustrates the basic equipment configuration used for these measurements. Fig. 5.51 shows this type of measuring equipment in use.

Depending on the type of measurement to be carried out, different acceleration sensors are used that are usually fixed to the object by magnets [5.5]. Fig. 5.52 shows an example of how acceleration sensors are placed on the production pallet of a circulation system.

Using the same principle, vibration measurements can also be carried out on any other vibration equipment used to manufacture concrete products. For instance, Fig. 5.53 shows the arrangement of acceleration sensors on the vibrating table of a block machine.

Fig. 5.53:
Arrangement of acceleration sensors on the vibrating table of a block machine

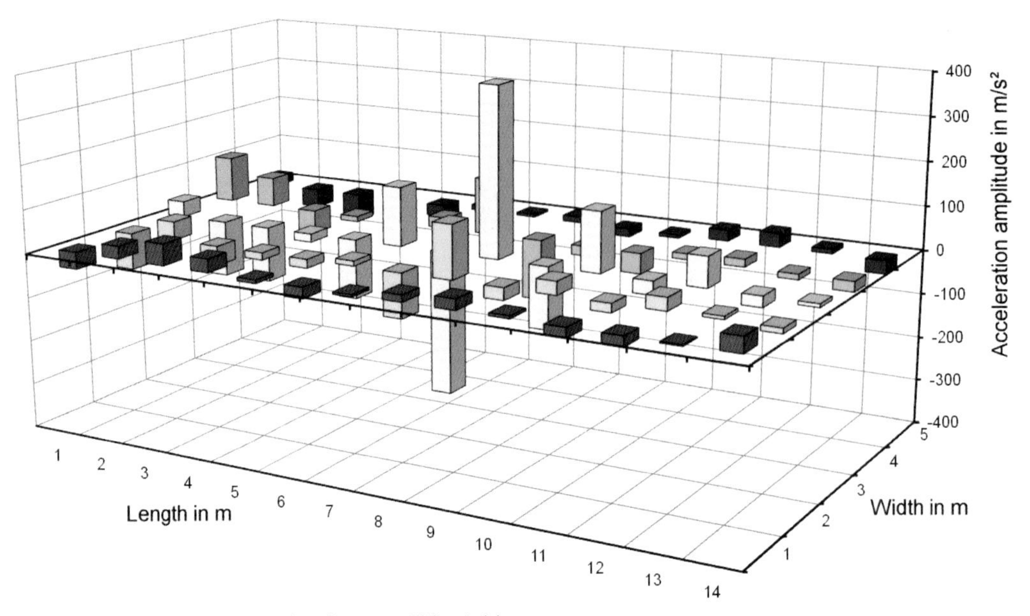

Fig. 5.54: Distribution of acceleration on a tilting table

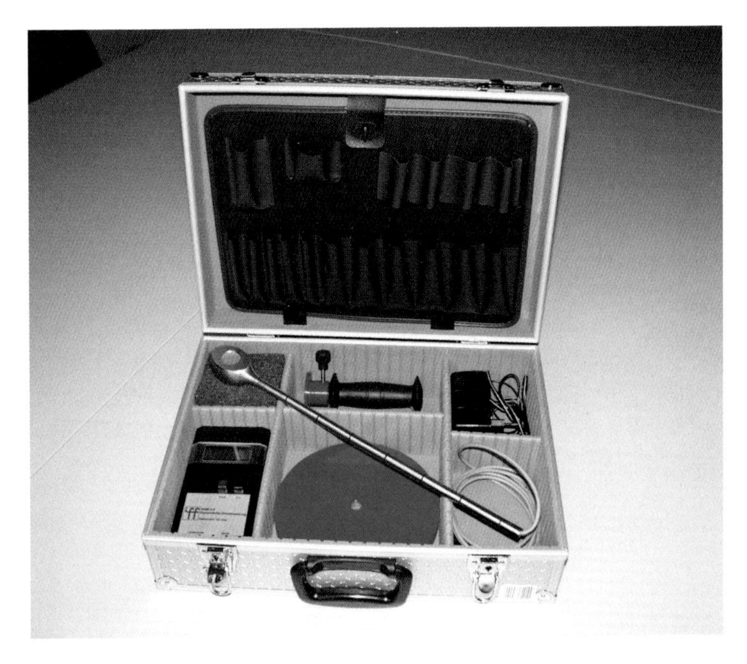

Fig. 5.55:
Measuring equipment

The results of such vibration measurements can be analysed in many different ways. Fig. 5.54 shows the distribution of acceleration measured on a tilt table.

Investigations referred to in [5.7] have shown that the dynamic change in pressure is a parameter well-suited to determining the vibration effect in plastic concrete mixes. For this purpose, a measuring device was developed whose pressure sensor is immersed in the mix in order to determine the dynamic changes in pressure that occur during the compaction process (Fig. 5.55). The sensor head captures changes in pressure as a three-dimensional pattern. Evaluation of the measuring device both at the pilot stage and in an industrial environment showed that the values measured for the changes in pressure within the mix largely corresponded to the achieved properties of the fresh and hardened concrete. These findings were used to develop an instrumentation and control concept for moulding and compaction of the concrete mix in the production of precast elements [5.7].

5.8 Selection Criteria

The following table lists several key criteria to be applied to the selection of equipment and machinery for producing precast elements [5.12].

1. Precast product	Which products are to be manufactured? - load-bearing solid walls - non-load bearing solid walls - double walls - partitions - facade panels - curtain walls - solid floors - filigree floors - noise barriers - other Which dimensions are to be produced? Which masses are to be moved/conveyed? What surface quality has been specified for the internal side of the wall? - mould surface - smoothed - levelled What surface quality has been specified for the external side of the wall? - levelled - smoothed - mould surface - no additional finishing - exposed-aggregate concrete - coloured - acid-washed - textured - pointed - painted - rendered

2. Concrete	Which concrete grade is to be processed? – normal concrete – lightweight concrete – heavyweight concrete – coloured concrete – no-fines concrete – self-compacting concrete – prestressed concrete How can the concrete mix be characterised? – workability (slump, w/c ratio) – mix design – shape of aggregates – round particles – crushed particles – other – maximum aggregate size Which properties should the fresh concrete have? – density – curing time
3. Reinforcement	Which reinforcement system has been specified? – individual rebars – lattice girders – mesh reinforcement – pre-fabricated reinforcement cages Is prestressed reinforcement required? Maximum weight of reinforcement unit?
4. Production equipment	Which equipment is appropriate to produce the required element shapes? Which production processes are to be used? Which type of moulding and compaction is to be considered? – self-compacting – vibrating – oscillating – shock vibration – combinations Which products should be manufactured in which period? Multi-shift operation? Is the intended location appropriate for the equipment? – building dimensions – ground conditions Can the production equipment achieve the required production output under the prevailing conditions? Have upstream and downstream equipment and processes been designed to achieve this output? – mixer – transport out of the factory – storage Which CAD/PPS system should be used?

5. Quality	Which parameters are key to the production process? – compaction parameters – acceleration amplitude – frequency – tolerances Which parameters should be achieved by the machine? Which options are required to influence the production parameters? Which in-process quality control measures exist or need to be implemented? Which quality control measures exist for upstream and downstream processes?
6. Flexibility	How are moulds exchanged? – time – cost Which mixes should or can be processed? Which product changeovers are planned, and how often? Can the production line be extended? Are breakdowns or emergencies possible? – Which? – Required actions?

6 Outlook

The increasingly diverse range of concrete products manufactured worldwide requires a flexible and quality-conscious approach to production. This situation imposes corresponding requirements both on the manufacturers of the associated production equipment and on its operators, i.e. precast plants. A large number of aspects need to be considered in this regard.

On the materials side, the ongoing development of ultra-high performance concrete (UHPC) and, in particular, the resource-saving use of building materials is crucial. For this reason, the use of recycled materials is one of the key factors to be considered. Another, similar factor is the use of renewable resources for fibre- and textile-reinforcement materials as well as the addition of other fibres and reinforcement materials to the concrete itself. This scenario results in new conditions as regards methods of processing these concrete mixes. Modelling and simulation of relevant process steps such as mixing, conveying, spreading, compacting and demoulding will become increasingly important. The use of new hardware and software will support this trend.

The design and engineering of production equipment will thus increasingly rely on dynamic modelling and simulation of equipment parameters. The use of engineering strength data will boost progress in determining the reliability and service life of the equipment.

It is generally necessary to use equipment that acts on the concrete mix so that the quality specifications of concrete products are consistently met in the production process.

In this respect, in-process quality control will become increasingly important as the basis for appropriate process instrumentation and control. Additional measuring equipment options will soon be available in this field. Solutions to mitigate noise and unwanted vibrations will gain in importance.

Existing handling and robot technology options will continue to be utilised and ultimately give rise to fully automated systems. Another relevant area is the development of equipment for manufacturing concrete products in very small quantities. Developing countries need tailored technical solutions to respond to specific conditions, such as the local availability of raw materials or the energy situation.

The increase in environmental awareness, the promotion of low-emission, energy-efficient processes and, in particular, the rise in energy costs have recently brought about a rethinking in the industry. In addition to the high quality of the processes, their efficiencies in terms of materials and energy are becoming increasingly important.

7 Bibliography

Chapter 1

[1.1] Kaysser, D.: Studie zur Analyse und Bewertung bekannter Fertigteilbausysteme hinsichtlich der Umweltrelevanz in der Phase der Fertigung. Forschungsbericht zum Projekt PRODOMO. Institut für Fertigteiltechnik und Fertigbau Weimar e.V., 1997

[1.2] Schwarz, S.: Flexible Umlaufproduktion mit angegliederten Vorbereitungsplätzen. Betonwerk+Fertigteil-Technik, Vol. 10/1994

[1.3] Autorenkollektiv: Handbuch Betonfertigteile, Betonwaren, Terrazzo. Verlag Bau+Technik GmbH, Düsseldorf 1999

[1.4] Kaysser, D.: Vibrationsverdichtung von Beton; grundlegende Sachverhalte und Zusammenhänge. Tagungsband Ulmer Beton- und Fertigteiltage, 1992

[1.5] Kuch, H.: Moderne Verfahren für die Formgebung und Verdichtung von Betonwaren und Betonfertigteilen [Modern procedures for the forming and compaction of concrete products and precast concrete elements]. Proceedings of the 15th International congress of the precast concrete industry, Paris 1996

[1.6] Afanasjew, A. A.: Technologie der Impulsverdichtung von Betongemengen. Moskau, Bauverlag 1986

[1.7] Kuch, H.: Modellbildung bei der Vibrationsverdichtung von Beton. Betonwerk+Fertigteiltechnik, Heft 2/1992

[1.8] Kuch, H.: Verfahrenstechnische Probleme bei der Formgebung und Verdichtung kleinformatiger Betonerzeugnisse. Betonwerk+Fertigteil-Technik, Heft 4/1992

[1.9] DIN 4235-3:1978-12 – Teil 3 [Part 3]: Verdichten von Beton durch Rütteln; Verdichten bei der Herstellung von Fertigteilen mit Außenrüttlern [Compacting of Concrete by Vibrating; Compacting by External Vibrators during the Manufacture of Precast Components]

[1.10] DIN EN 197-1:2004-08 – Zement – Teil 1 [Cement – Part 1]: Zusammensetzung, Anforderungen und Konformitätskriterien von Normalzement [Composition, specifications and conformity criteria for common cements]

[1.11] DIN 1164-10:2004-08 – Zement mit besonderen Eigenschaften – Teil 10 [Special cement – Part 10]: Zusammensetzung, Anforderungen und Übereinstimmungsnachweis von Normalzement mit besonderen Eigenschaften [Composition, requirements and conformity evaluation for special common cement]

[1.12] DIN 1164-11:2003-11 – Zement mit besonderen Eigenschaften – Teil 11 [Special cement – Part 11]: Zusammensetzung, Anforderungen und Übereinstimmungsnachweis von Zement mit verkürztem Erstarren [Composition, specification and conformity evaluation for cement with short setting time]

[1.13] DIN 1164-12:2005-06 – Zement mit besonderen Eigenschaften – Teil 12 [Special cement – Part 12]: Zusammensetzung, Anforderungen und Übereinstimmungsnachweis von Zement mit einem erhöhten Anteil an organischen Bestandteilen [Composition, specification and conformity evaluation for cement with higher quantity of organic constituents]

[1.14] DIN EN 12620:2008-07 – Gesteinskörnungen für Beton [Aggregates for concrete]

[1.15] DIN EN 13055-1:2002-08 – Leichte Gesteinskörnungen – Teil 1 [Lightweight aggregates - Part 1]: Leichte Gesteinskörnungen für Beton, Mörtel und Einpressmörtel [Lightweight aggregates for concrete, mortar and grout]
[1.16] DIN EN 4226-100:2002-02 – Gesteinskörnungen für Beton und Mörtel – Teil 100 [Aggregates for concrete and mortar - Part 100]: Rezyklierte Gesteinskörnungen [Recycled aggregates]
[1.17] DIN 1045-2:2008-08 – Tragwerke aus Beton, Stahlbeton und Spannbeton – Teil 2 [Concrete, reinforced and prestressed concrete structures - Part 2]: Beton-Festlegung, Eigenschaften, Herstellung und Konformität – Anwendungsregeln zu DIN EN 206-1 [Concrete - Specification, properties, production and conformity - Application rules for DIN EN 206-1]
[1.18] DIN EN 206-1:2001-07 – Beton – Teil 1 [Concrete – Part 1]: Festlegung, Eigenschaften, Herstellung und Konformität [Specification, performance, production and conformity]
[1.19] DIN EN 12878:2006-05 – Pigmente zum Einfärben von zement- und/oder kalkgebundenen Baustoffen – Anforderungen und Prüfverfahren [Pigments for the colouring of building materials based on cement and/or lime – Specifications and methods of test]
[1.20] DIN 51043:1979-08 – Trass; Anforderungen, Prüfung [Trass; Requirements, Tests]
[1.21] DIN EN 450-1:2008-05 – Flugasche für Beton – Teil 1 [Fly ash for concrete – Part 1]: Definition, Anforderungen und Konformitätskriterien [Definition, specifications and conformity criteria]
[1.22] DIN EN 13263-1:2009-07 – Silikastaub für Beton – Teil 1 [Silica fume for concrete - Part 1]: Definitionen, Anforderungen und Konformitätskriterien [Definitions, requirements and conformity criteria]
[1.23] DIN EN 14889-1:2006-11 – Fasern für Beton – Teil 1 [Fibres for concrete - Part 1]: Stahlfasern - Begriffe, Festlegungen und Konformität [Steel fibres - Definitions, specifications and conformity]
[1.24] DIN EN 14889-2:2006-11 – Fasern für Beton – Teil 2 [Fibres for concrete - Part 2]: Polymerfasern – Begriffe, Festlegungen und Konformität [Polymer fibres - Definitions, specifications and conformity]
[1.25] DIN EN 1008:2002-10 – Zugabewasser für Beton – Festlegung für die Probenahme, Prüfung und Beurteilung der Eignung von Wasser, einschließlich bei der Betonherstellung anfallendem Wasser, als Zugabewasser für Beton [Mixing water for concrete - Specification for sampling, testing and assessing the suitability of water, including water recovered from processes in the concrete industry, as mixing water for concrete]
[1.26] Walz, K.: Beziehung zwischen Wasserzementwert, Normfestigkeit des Zements (DIN 1164, June 1970) und Betondruckfestigkeit. In: Beton 20 (1970) 11, 499-503
[1.27] Wesche, K.: Baustoffe für tragende Bauteile: Band 2: Beton, Mauerwerk (Nichtmetallisch-anorganische Stoffe); Herstellung, Eigenschaften, Verwendung, Dauerhaftigkeit. 3. völlig neu-bearb. u. erw. Aufl. Bauverlag GmbH, Wiesbaden 1993
[1.28] Betontechnische Daten. Ed. Heidelberg Cement AG. 2008 edition

[1.29] Verein Deutscher Zementwerke e. V.: Zement Taschenbuch 51. Ausgabe. Verlag Bau+Technik GmbH, Düsseldorf 2008

[1.30] Dickamp. J. Richter, T.: Frischbeton, Eigenschaften und Prüfung. Zementmerkblatt Betontechnik B4, 1.2007. Ed.: Verein deutscher Zementwerke e. V.

[1.31] Wierig, H.-J.: Eigenschaften von grünem, jungem Beton: Druckfestigkeit, Verformungsverhalten, Wasserverdunstung. In: Beton 18 (1968) 3, 94-101

[1.32] Stark, J.; Wicht, B.: Dauerhaftigkeit von Beton: Der Baustoff als Werkstoff. Birkhäuser Verlag, Basel 2001

[1.33] Eifert, H.; Bethge, W.: Beton – Prüfung nach Norm: Die neue Normengeneration. Verlag Bau+Technik GmbH, Düsseldorf 2005

[1.34] DIN EN 12350-4:2000-06 – Prüfung von Frischbeton – Teil 4 [Testing fresh concrete - Part 4]: Verdichtungsmaß [Degree of compactability]

[1.35] DIN EN 12350-5:2000-06 – Prüfung von Frischbeton – Teil 5 [Testing fresh concrete - Part 5]: Ausbreitmaß [Flow table test]

[1.36] DIN EN 12350-7: 2000-11 – Prüfung von Frischbeton – Teil 7 [Testing fresh concrete - Part 7]: Luftgehalte, Druckverfahren [Air content - Pressure methods]

[1.37] DIN 18127:1997-11 – Baugrund – Untersuchung von Bodenproben – Proctorversuch [Soil; Investigation and testing – Proctor test]

[1.38] DIN EN 1015-12:2000-06 – Prüfverfahren für Mörtel für Mauerwerk – Teil 12 [Methods of test for mortar for masonry - Part 12]: Bestimmung der Haftzugfestigkeit von erhärtetem Putzmörtel [Determination of adhesive strength of hardened rendering and plastering mortars on substrates]

[1.39] DIN EN 12390-8:2001-02 – Prüfung von Festbeton – Teil 8 [Testing hardened concrete - Part 8]: Wassereindringtiefe unter Druck [Depth of penetration of water under pressure]

[1.40] DIN 1048-5:1991-06 – Prüfverfahren für Beton; Festbeton, gesondert hergestellte Probekörper [Testing concrete; testing of hardened concrete (specimens prepared in mould)]

[1.41] Mechtcherine, V., Götze, M.: Institut für Baustoffe, TU Dresden, ICCX (International Concrete Conference & Exhibition) 2008, St. Petersburg, 9–11 December 2008, Erdfeuchter Beton – Grundlagen, Anwendung und Optimierung [Developing Recipes for Zero Slump Concretes]

[1.42] Bornemann, R.: Untersuchungen zur Modellierung des Frisch- und Festbetonverhaltens erdfeuchter Betone. Dissertation. Universität Kassel. Schriftenreihe Baustoffe und Massivbau, Vol. 4, 2005

[1.43] Momber, A. W.; Schulz, R. -R.: Handbuch der Oberflächenbearbeitung Beton, Birkhäuser Verlag, Basel 2006

[1.44] DIN EN 13813:2003-01 – Estrichmörtel, Estrichmassen und Estriche – Estrichmörtel und Estrichmassen – Eigenschaften und Anforderungen [Screed material and floor screeds - Screed materials - Properties and requirements]

[1.45] DAfStb-Richtlinie: Schutz und Instandsetzung von Betonbauteilen (Instandsetzungs-Richtlinie) – Teil 2: Bauprodukte und Anwendung. October 2001. Ed.: Deutscher Ausschuss für Stahlbeton (DAfStb)

[1.46] DIN 1045-1:2008-08 – Tragwerke aus Beton, Stahlbeton und Spannbeton – Teil 1 [Concrete, reinforced and prestressed concrete structures - Part 1]: Bemessung und Konstruktion [Design and construction]

[1.47] DIN EN 12504-4:2004-12 – Prüfung von Beton in Bauwerken – Teil 4 [Testing concrete in structures - Part 4]: Bestimmung der Ultraschallgeschwindigkeit [Determination of ultrasonic pulse velocity]

[1.48] Bunke, N.: Prüfung von Beton. Empfehlungen und Hinweise als Ergänzung zu DIN 1048. Deutscher Ausschuss für Stahlbeton (DAfStb) Vol. 422, Beuth-Verlag GmbH, Berlin 1991

[1.49] DIN CEN/TS 12390-9:2006-08 – Prüfung von Festbeton – Teil 9 [Testing hardened concrete - Part 9]: Frost- und Frost-Tausalz-Widerstand; Abwitterung [Freeze-thaw resistance – Scaling]

[1.50] DAfStb-Richtlinie Vorbeugende Maßnahmen gegen schädigende Alkalireaktion im Beton (Alkali-Richtlinie), February 2007, Ed.: Deutscher Ausschuss für Stahlbeton (DAfStb)

[1.51] Pflastersteine aus Beton nach neuer europäischer Norm DIN EN 1338. Informationen für Planer, Ausführende, Baustoffhandel und Bauherren, Ed.: Betonverband Straße, Landschaft, Garten e. V. (SLG), updated version, December 2007

[1.52] TL-Pflaster-StB 06 – Technische Lieferbedingungen für Bauprodukte zur Herstellung von Pflasterdecken, Plattenbelägen und Einfassungen, 2006 edition, FGSV-Verlag, Cologne 2006

[1.53] BGB-Richtlinie Nicht genormte Betonprodukte – Anforderungen und Prüfungen – (BGB-RiNGB), 2005 edition

[1.54] Platten aus Beton nach neuer europäischer Norm DIN EN 1339. Informationen für Planer, Ausführende, Baustoffhandel und Bauherren, Ed.: Betonverband Straße, Landschaft, Garten e.V. (SLG), updated version, December 2007

[1.55] Bordsteine aus Beton nach neuer europäischer Norm DIN EN 1340. Informationen für Planer, Ausführende, Baustoffhandel und Bauherren, Ed.: Betonverband Straße, Landschaft, Garten e.V. (SLG), updated version, December 2007

[1.56] FBS-Qualitätsrichtlinie Betonrohre, Stahlbetonrohre und Vortriebsrohre mit Kreisquerschnitt in FBS-Qualität für erdverlegte Abwasserleitungen und -kanäle, Ausführungen, Anforderungen und Prüfungen. Teil 1-1, August 2005, Ed.: Fachvereinigung Betonrohre und Stahlbetonrohre

[1.57] FBS-Qualitätsrichtlinie Betonrohre, Stahlbetonrohre und Vortriebsrohre mit Kreisquerschnitt in FBS-Qualität für erdverlegte Abwasserleitungen und -kanäle, Ausführungen, Anforderungen und Prüfungen. Teil 1-1, August 2005, Ed.: Fachvereinigung Betonrohre und Stahlbetonrohre

[1.58] FBS-Qualitätsrichtlinie Betonrohre, Stahlbetonrohre und Vortriebsrohre mit Eiquerschnitt in FBS-Qualität für erdverlegte Abwasserleitungen und -kanäle, Ausführungen, Anforderungen und Prüfungen. Teil 1-2, August 2005, Ed.: Fachvereinigung Betonrohre und Stahlbetonrohre

[1.59] FBS-Qualitätsrichtlinie Sonderquerschnitte und Sonderausführungen von Betonrohren und Stahlbetonrohren in FBS-Qualität für erdverlegte Abwasserleitungen und -kanäle, Ausführungen, Anforderungen und Prüfungen. Teil 1-3, August 2005, Ed.: Fachvereinigung Betonrohre und Stahlbetonrohre

[1.60] FBS-Qualitätsrichtlinie Formstücke aus Beton und Stahlbeton in FBS-Qualität für erdverlegte Abwasserleitungen und -kanäle, Ausführungen, Anforderungen und Prüfungen, Teil 1-4, August 2005, Ed.: Fachvereinigung Betonrohre und Stahlbetonrohre

[1.61] FBS-Qualitätsrichtlinie Betonrohre und Stahlbetonrohre mit Zuläufen (Abzweigen) in FBS-Qualität für erdverlegte Abwasserleitungen und -kanäle, Ausführungen, Anforderungen und Prüfungen. Teil 1-5, August 2005, Ed.: Fachvereinigung Betonrohre und Stahlbetonrohre

[1.62] FBS-Qualitätsrichtlinie Rohre und Formstücke aus Beton und Stahlbeton für erdverlegte Abwasserleitungen und -kanäle, Ausführungen, Anforderungen und Prüfungen. Teil 2, August 2005, Ed.: Fachvereinigung Betonrohre und Stahlbetonrohre

[1.63] FBS-Qualitätsrichtlinie Schachtfertigteile aus Beton und Stahlbeton und Schachtbauwerke aus Stahlbetonfertigteilen für erdverlegte Abwasserleitungen und -kanäle, Ausführungen, Anforderungen und Prüfungen. Teil 2-1, August 2005, Ed.: Fachvereinigung Betonrohre und Stahlbetonrohre

[1.64] FBS-Qualitätsrichtlinie Schachtbauwerke aus Stahlbetonfertigteilen für erdverlegte Abwasserleitungen und -kanäle, Ausführungen, Anforderungen und Prüfungen. Teil 2-2, August 2005, Ed.: Fachvereinigung Betonrohre und Stahlbetonrohre

[1.65] DIN EN 1338:2003-08 Pflastersteine aus Beton – Anforderungen und Prüfverfahren [Concrete paving blocks - Requirements and test methods]

[1.66] DIN EN 1339:2003-08 Platten aus Beton – Anforderungen und Prüfverfahren [Concrete paving flags - Requirements and test methods]

[1.67] DIN EN 1340:2003-08 Bordsteine aus Beton – Anforderungen und Prüfverfahren [Concrete kerb units; Requirements and test methods]

[1.68] DIN EN 13748-2: 2005-03 – Terrazzoplatten – Teil 2 [Terrazzo tiles - Part 2]: Terrazzoplatten für die Verwendung im Außenbereich [Terrazzo tiles for external use]

[1.69] DIN EN 771-3:2005-05 Festlegungen für Mauersteine – Teil 3 [Specification for masonry units - Part 3]: Mauersteine aus Beton (mit dichten und porigen Gesteinskörnungen) [Aggregate concrete masonry units (Dense and light-weight aggregates)]

[1.70] DIN V 18151-100:2005-10 – Hohlblöcke aus Leichtbeton – Teil 100 [Lightweight concrete hollow blocks - Part 100]: Hohlblöcke mit besonderen Eigenschaften [Hollow blocks with specific properties]

[1.71] DIN V 18152-100:2005-10 – Vollsteine und Vollblöcke aus Leichtbeton – Teil 100 [Lightweight concrete solid bricks and blocks - Part 100]: Vollsteine und Vollblöcke mit besonderen Eigenschaften [Solid bricks and blocks with specific properties]

[1.72] DIN V 18153-100:2005-10 – Mauersteine aus Beton (Normalbeton) – Teil 100 [Concrete masonry units (Normal-weight concrete) - Part 100]: Mauersteine mit besonderen Eigenschaften [Masonry units with specific properties]

[1.73] DIN EN 490:2005-03 – Dach- und Formsteine aus Beton für Dächer und Wandbekleidungen – Produktanforderungen [Concrete roofing tiles and fittings for roof covering and wall cladding - Product specifications]

[1.74] DIN EN 491:2005-03 – Dach- und Formsteine aus Beton für Dächer und Wandbekleidungen – Prüfverfahren [Concrete roofing tiles and fittings for roof covering and wall cladding - Test methods]

[1.75] DIN V 18500:2006-12 – Betonwerkstein – Begriffe, Anforderungen, Prüfung, Überwachung [Cast stones - Terminology, requirements, testing, inspection]
[1.76] DIN EN 13369:2004-09 – Allgemeine Regeln für Betonfertigteile [Common rules for precast concrete products]
[1.77] DIN EN 1916:2003-04 – Rohre und Formstücke aus Beton, Stahlfaserbeton und Stahlbeton [Concrete pipes and fittings, unreinforced, steel fibre and reinforced]
[1.78] DIN V 1201:2004-08 – Rohre und Formstücke aus Beton, Stahlfaserbeton und Stahlbeton für Abwasserleitungen und -kanäle – Typ 1 und Typ 2 – Anforderungen, Prüfung und Bewertung der Konformität [Concrete pipes and fittings, unreinforced, steel fibre and reinforced for drains and sewers - Type 1 and Type 2 - Requirements, test methods and evaluation of conformity]
[1.79] DIN EN 1917:2003-04 – Einsteig- und Kontrollschächte aus Beton, Stahlfaserbeton und Stahlbeton [Concrete manholes and inspection chambers, unreinforced, steel fibre and reinforced]
[1.80] DIN V 4034-1:2004-08 – Schächte aus Beton-, Stahlfaserbeton- und Stahlbetonfertigteilen für Abwasserleitungen und -kanäle – Typ 1 und Typ 2 – Teil 1 [Prefabricated concrete manholes, unreinforced, steel fibre and reinforced for drains and sewers - Type 1 and Type 2 - Part 1]: Anforderungen, Prüfung und Bewertung der Konformität [Requirements, test methods and evaluation of conformity]
[1.81] BGB-Richtlinie Nicht genormte Betonprodukte – Anforderungen und Prüfungen – (BGB-RiNGB), Ed. Bund Güteschutz Beton- und Stahlbetonfertigteile e.V., 2005 edition
[1.82] DIN EN 12390-2:2001-06 – Prüfung von Festbeton – Teil 2 [Testing hardened concrete - Part 2]: Herstellung und Lagerung von Probekörpern für Festigkeitsprüfungen [Making and curing specimens for strength tests]
[1.83] DIN EN 12390-3:2009-07 – Prüfung von Festbeton – Teil 3 [Testing hardened concrete - Part 3]: Druckfestigkeit von Probekörpern [Compressive strength of test specimens]
[1.84] DIN EN 12390-4:2000-12 – Prüfung von Festbeton – Teil 4 [Testing hardened concrete - Part 4]: Bestimmung der Druckfestigkeit; Anforderungen an Prüfmaschinen [Compressive strength; Specification for testing machines]
[1.85] DIN EN 12390-5:2009-07 – Prüfung von Festbeton – Teil 5 [Testing hardened concrete - Part 5]: Biegezugfestigkeit von Probekörpern [Flexural strength of test specimens]
[1.86] DIN EN 12390-6:2001-02 – Prüfung von Festbeton – Teil 6 [Testing hardened concrete - Part 6]: Spaltzugfestigkeit von Probekörpern [Tensile splitting strength of test specimens]
[1.87] DIN EN 12390-7:2009-07 – Prüfung von Festbeton – Teil 7 [Testing hardened concrete - Part 7]: Dichte von Festbeton [Density of hardened concrete]
[1.88] DIN EN 12504-1:2000-09 – Prüfung von Beton in Bauwerken – Teil 1 [Testing concrete in structures - Part 1]: Bohrkernproben – Herstellung, Untersuchung und Prüfung der Druckfestigkeit [Cored specimens - Taking, examining and testing in compression]
[1.89] DIN EN ISO 140-3:2005-03 – Akustik – Messung der Schalldämmung in Gebäuden und von Bauteilen – Teil 3 [Acoustics - Measurement of sound insulation in

buildings and of building elements - Part 3]: Messung der Luftschalldämmung von Bauteilen in Prüfständen [Laboratory measurements of airborne sound insulation of building elements]

[1.90] DIN EN ISO 140-6:1998-12 – Akustik – Messung der Schalldämmung in Gebäuden und von Bauteilen – Teil 6 [Acoustics - Measurement of sound insulation in buildings and of building elements - Part 6]: Messung der Trittschalldämmung von Decken in Prüfständen [Laboratory measurements of impact sound insulation of floors]

[1.91] DIN EN 12664:2001-05 – Wärmetechnisches Verhalten von Baustoffen und Bauprodukten – Bestimmung des Wärmedurchlasswiderstands nach dem Verfahren mit dem Plattengerät und dem Wärmestrommessplatten-Gerät – Trockene und feuchte Produkte mit mittlerem und niedrigem Wärmedurchlasswiderstand [Thermal performance of building materials and products - Determination of thermal resistance by means of guarded hot plate and heat flow meter methods - Dry and moist products with medium and low thermal resistance]

[1.92] DIN EN ISO 6946:2008-04 – Bauteile – Wärmedurchlasswiderstand und Wärmedurchgangskoeffizient – Berechnungsverfahren [Building components and building elements - Thermal resistance and thermal transmittance - Calculation method]

[1.93] DIN EN ISO 8990:1996-09 – Wärmeschutz – Bestimmung der Wärmedurchgangseigenschaften im stationären Zustand – Verfahren mit dem kalibrierten und dem geregelten Heizkasten [Thermal insulation - Determination of steady-state thermal transmission properties - Calibrated and guarded hot box]

[1.94] DIN EN 1934:1998-04 – Wärmetechnisches Verhalten von Gebäuden – Messung des Wärmedurchlasswiderstands; Heizkastenverfahren mit dem Wärmestrommesser – Mauerwerk [Thermal performance of buildings - Determination of thermal resistance by hot box method using heat flow meter – Masonry]

[1.95] DIN 1048-5:1991-06 – Prüfverfahren für Beton; Festbeton, gesondert hergestellte Probekörper [Testing concrete; testing of hardened concrete (specimens prepared in mould)]

[1.96] DIN 1048-2:1991-06 – Prüfverfahren für Beton; Festbeton in Bauwerken und Bauteilen [Testing concrete; testing of hardened concrete (specimens taken in situ)]

[1.97] DIN EN 771-3:2005-05 - Festlegungen für Mauersteine – Teil 3 [Specification for masonry units - Part 3]: Mauersteine aus Beton (mit dichten und porigen Gesteinskörnungen) [Aggregate concrete masonry units (Dense and light-weight aggregates)]

[1.98] DIN 1053-2:1996-11 – Mauerwerk – Teil 2 [Masonry - Part 2]: Mauerwerksfestigkeitsklassen aufgrund von Eignungsprüfungen [Masonry strength classes on the basis of suitability tests]

[1.99] DIN V 18151-100:2005-10 – Hohlblöcke aus Leichtbeton – Teil 100 [Lightweight concrete hollow blocks - Part 100]: Hohlblöcke mit besonderen Eigenschaften [Hollow blocks with specific properties]

[1.100] DIN V 18152-100:2005-10 – Vollsteine und Vollblöcke aus Leichtbeton – Teil 100 [Lightweight concrete solid bricks and blocks - Part 100]: Vollsteine und Vollblöcke mit besonderen Eigenschaften [Solid bricks and blocks with specific properties]

[1.101] DIN V 18153-100:2005-10 – Mauersteine aus Beton (Normalbeton) – Teil 100 [Concrete masonry units (Normal-weight concrete) - Part 100]: Mauersteine mit besonderen Eigenschaften [Masonry units with specific properties]

[1.102] DIN EN 772-1:2000-09 – Prüfverfahren für Mauersteine – Teil 1 [Methods of test for masonry units - Part 1]: Bestimmung der Druckfestigkeit [Determination of compressive strength]

[1.103] DIN EN 772-16:2005-05 – Prüfverfahren für Mauersteine – Teil 16 [Methods of test for masonry units - Part 16]: Bestimmung der Maße [Determination of dimensions]

[1.104] DIN EN 772-2:2005-05 – Prüfverfahren für Mauersteine – Teil 2 [Methods of test for masonry units - Part 2]: Bestimmung des prozentualen Lochanteils in Mauersteinen (mittels Papiereindruck) [Determination of percentage area of voids in masonry units (by paper indentation)]

[1.105] DIN EN 772-20:2005-05 – Prüfverfahren für Mauersteine – Teil 20 [Methods of test for masonry units - Part 20]: Bestimmung der Ebenheit von Mauersteinen [Determination of flatness of faces of masonry units]

[1.106] DIN EN 772-1:2000-09 – Prüfverfahren für Mauersteine – Teil 1 [Methods of test for masonry units - Part 1]: Bestimmung der Druckfestigkeit [Determination of compressive strength]

[1.107] DIN EN 772-6:2002-02 – Prüfverfahren für Mauersteine – Teil 6 [Methods of test for masonry units - Part 6]: Bestimmung der Biegezugfestigkeit von Mauersteinen aus Beton [Determination of bending tensile strength of aggregate concrete masonry units]

[1.108] DIN EN 772-14:2002-02 – Prüfverfahren für Mauersteine – Teil 14 [Methods of test for masonry units - Part 14]: Bestimmung der feuchtebedingten Formänderung von Mauersteinen aus Beton und Betonwerksteinen [Determination of moisture movement of aggregate concrete and manufactured stone masonry units]

[1.109] DIN EN 772-3:1998-10 – Prüfverfahren für Mauersteine – Teil 3 [Methods of test for masonry units - Part 3]: Bestimmung des Nettovolumens und des prozentualen Lochanteils von Mauerziegeln mittels hydrostatischer Wägung (Unterwasserwägung) [Determination of net volume and percentage of voids of clay masonry units by hydrostatic weighing]

[1.110] DIN EN 772-11:2004-06 – Prüfverfahren für Mauersteine – Teil 11 [Methods of test for masonry units – Part 11]: Bestimmung der kapillaren Wasseraufnahme von Mauersteinen aus Beton, Porenbetonsteinen, Betonwerksteinen und Natursteinen sowie der anfänglichen Wasseraufnahme von Mauerziegeln [Determination of water absorption of aggregate concrete, autoclaved aerated concrete, manufactured stone and natural stone masonry units due to capillary action and the initial rate of water absorption of clay masonry units]

[1.111] DIN EN 772-13:2000-09 – Prüfverfahren für Mauersteine – Teil 13 [Methods of test for masonry units - Part 13]: Bestimmung der Netto- und Brutto-Trockenrohdichte von Mauersteinen (außer Natursteinen) [Determination of net and gross dry density of masonry units (except for natural stone)]

[1.112] DIN EN 1052-3:2007-06 – Prüfverfahren für Mauerwerk – Teil 3 [Methods of test for masonry - Part 3]: Bestimmung der Anfangsscherfestigkeit (Haftscherfestigkeit) [Determination of initial shear strength]

[1.113] DIN EN 13501-1:2007-05 – Klassifizierung von Bauprodukten und Bauarten zu ihrem Brandverhalten – Teil 1 [Fire classification of construction products and building elements - Part 1]: Klassifizierung mit den Ergebnissen aus den Prüfungen zum Brandverhalten von Bauprodukten [Classification using data from reaction to fire tests]

[1.114] DIN EN ISO 12572:2001-09 – Wärme- und feuchtetechnisches Verhalten von Baustoffen und Bauprodukten – Bestimmung der Wasserdampfdurchlässigkeit [Hygrothermal performance of building materials and products - Determination of water vapour transmission properties]

[1.115] DIN EN 1745:2002-08 – Mauerwerk und Mauerwerksprodukte – Verfahren zur Ermittlung von Wärmeschutzrechenwerten [Masonry and masonry products - Methods for determining design thermal values]

[1.116] Kuch, H.; Walter, M.; Schwabe, J.-H.: Einflussgrößen auf die qualitätsgerechte Fertigung von Betonwaren. In: BetonWerk International, Vol. 05/2004, 106-112

[1.117] Kuch, H.; Schwabe, J.-H.: Optimierung der Verarbeitungsprozesse von Betongemengen zur Erzielung höherer Qualität von Betonwaren. Presentation at the 50th BetonTage congress, 14-16 February 2006, Ulm. In: Betonwerk+Fertigteil-Technik Vol. 02/2006, 44-46

[1.118] Kuch, H.; Schwabe, J.-H.: Development and control of concrete mix processing procedures. In: Proceedings of the 18th BIBM International Congress. 11-14 May 2005, Amsterdam, 108-109

[1.119] Kuch, H.; Palzer, S.; Schwabe, J.-H.: Anwendung der Simulation bei der Verarbeitung von Gemengen. Tagungsbericht, Vol. 1, 1-1321 to 1-1327, 16. Internationale Baustofftagung 2006, Weimar

[1.120] Martin, M.; Schulze, R.: Grundlagen der Betonverdichtung. Wacker Construction Equipment AG, Munich 2008

[1.121] DIN 18200:2005-05 – Übereinstimmungsnachweis für Bauprodukte – Werkseigene Produktionskontrolle, Fremdüberwachung und Zertifizierung von Produkten [Assessment of conformity for construction products - Certification of construction products by certification body]

[1.122] DIN 1045-4:2001-07 – Tragwerke aus Beton, Stahlbeton und Spannbeton – Teil 4 [Concrete, reinforced and prestressed concrete structures - Part 4]: Ergänzende Regeln für die Herstellung und die Konformität von Fertigteilen [Additional rules for the production and conformity control of prefabricated elements]

[1.123] DIN 52108:2002-07 – Verschleißprüfung mit der Schleifmaschine nach Böhme [Wear test using the grinding wheel according to Böhme]

Chapter 2

[2.1] Beitzel, H.: Herstellen und Verarbeiten von Beton, Betonkalender 2003. Verlag Ernst & Sohn, Berlin 2003

[2.2] DIN 459-1, Mischer für Beton und Mörtel, Teil 1 [Building material machines - Mixers for concrete and mortar - Part 1]: Begriffe, Leistungsermittlung, Größen [Terms, determination of performance, sizes]

[2.3] DIN 459-2:1995-11 Mischer für Beton und Mörtel, Teil 2 [Building material machines - Mixers for concrete and mortar - Part 2]: Verfahren zur Prüfung der Mischwirkung von Betonmischern [Procedure for the examination of the mixing efficiency of concrete mixers]

[2.4] ISO 18650-2:2006-04 Building construction machinery and equipment – Concrete mixers, Part 2: Procedure for examination of mixing efficiency

Chapter 3

[3.1] Autorenkollektiv: Betonfertigteile, Betonwerkstein, Terrazzo. Handbuch. Verlag Bau+Technik GmbH, Düsseldorf 1999

[3.2] Mothes, St.: Die Füllung der Form mit Betongemenge bei der Formgebung und Verdichtung von Betonsteinen in Steinformmaschinen. Dissertation, Bauhaus-Universität Weimar, 2009

[3.3] DIN V 18500:2006-12 Betonwerkstein – Begriffe, Anforderungen, Prüfung, Überwachung, [Cast stones - Terminology, requirements, testing, inspection]

[3.4] DIN EN 490:2006-09 Dach- und Formsteine aus Beton für Dächer und Wandbekleidungen, Produktanforderungen; Deutsche Fassung EN 490:2004 + A1:2006 [Concrete roofing tiles and fittings for roof covering and wall cladding - Product specifications; German version EN 490:2004 + A1:2006]

[3.5] DIN EN 12629-2:2003-06 Maschinen für die Herstellung von Bauprodukten aus Beton- und Kalksandsteinmassen – Sicherheit; Teil 2 [Machines for the manufacture of constructional products from concrete and calcium-silicate - Safety - Part 2]: Steinformmaschinen [Block making machines]

[3.6] Kuch, H.; u. a.: Effektivierung der Auflastwirkung in Betonsteinfertigern. Schriftenreihe der Forschungsvereinigung Bau- und Baustoffmaschinen, December 2005 issue; Forschungsstelle: Institut für Fertigteiltechnik und Fertigbau Weimar e.V.

[3.7] Kuch, H.; u. a.: Schockvibrationsregime. Schriftenreihe der Forschungsvereinigung Bau- und Baustoffmaschinen, Vol. 14, June 1999; Forschungsstelle: Institut für Fertigteiltechnik und Fertigbau Weimar e.V.

[3.8] Mothes, St.: Erfahrungen mit der Harmonischen Vibration bei der Herstellung von Betonwaren. Betonwerk Informationen 06/2007, 90-97

[3.9] Schlecht, B.; Neubauer, A.: Steigerung der Produktqualität durch effiziente Verdichtung; Betonwerk+Fertigteil-Technik 09/2000, 44-52

[3.10] Schwabe, J.-H.; Kuch, H.; Mothes, S.: Harmonische Vibration bei Steinformmaschinen. Die Industrie der Steine+ Erden 01/2006, 30-34

[3.11] Autorenkollektiv: Weiterentwicklung des Füllprozesses in Betonsteinfertigern zur Gewährleistung homogener Produktqualität von Betonwaren. Auftraggeber Bundesministerium für Forschung und Technologie Berlin. Auftragnehmer: Institut für Fertigteiltechnik und Fertigbau Weimar e. V.; 2007

[3.12] Kuch, H. et.al.: Einfluss der Fundamentierung bei der Lärm- und Schwingungsabwehr an Betonsteinfertigern. Schriftenreihe der Forschungsvereinigung Bau- und Baustoffmaschinen, Vol. 8, November 1996; Forschungsstelle: Institut für Fertigteiltechnik und Fertigbau Weimar e.V.
[3.13] Autorenkollektiv: Entwicklung eines industrietauglichen Verfahrens zur Qualitätssicherung von Betonsteinen durch Online-Überwachung des Verdichtungsprozesses anhand der Frequenzspektren der Beschleunigungen an signifikanten Maschinenpunkten. Forschungsprojekt in Auftraggeberschaft des Bundesministeriums für Forschung und Technologie; Auftragnehmer Institut für Fertigteiltechnik und Fertigbau Weimar e.V., 2001
[3.14] Autorenkollektiv: Entwicklung eines industrietauglichen Systems für das Erfassen qualitätsrelevanter Prozess- und Produktparameter bei der Herstellung von Betonwaren. Forschungsverbundprojekt: Auftraggeber: Thüringer Ministerium für Wissenschaft, Forschung und Kunst/Thüringer Ministerium für Wirtschaft, Technolgie und Arbeit.; Auftragnehmer: Institut für Fertigteiltechnik und Fertigbau Weimar e.V., 2006
[3.15] Schweyer, P.: Veredlung von Betonwaren – Beton ist, was man draus macht. 14. Fachtagung des IFF Weimar e.V., 2007, Tagungsband
[3.16] DIN EN 490:2006-09; Dach und Formsteine aus Beton für Dächer und Wandbekleidungen – Produktanforderungen; Deutsche Fassung EN 490:2004+ A1:2006 [Concrete roofing tiles and fittings for roof covering and wall cladding - Product specifications; German version EN 490:2004 + A1:2006]
[3.17] DIN EN 491:2005-03; Dach- und Formsteine aus Beton für Dächer und Wandbekleidungen – Prüfverfahren; Deutsche Fassung EN 491:2004 [Concrete roofing tiles and fittings for roof covering and wall cladding - Test methods; German version EN 491:2004]

Chapter 4

[4.1] Baumgärtner, G.: Das Rotationspressverfahren zur Herstellung von Betonrohren. Munich, Technische Universität; IFF Weimar; Diplomarbeit, 1997
[4.2] Kuch, H.; Schwabe, J.-H.: Aktueller Stand der Herstellung von Beton- und Stahlbetonrohren. Weimar. IFF Weimar e.V., 1994 – Forschungsbericht im Auftrag des Bayerischen Industrieverbands Steine und Erden
[4.3] Kuch, H.; Schwabe, J.-H.: Schwingungstechnische Modellierung und Berechnung der Verdichtungseinrichtungen zur Rohrherstellung. In: Betonwerk+Fertigteil-Technik Vol. 09/1996, 84-87
[4.4] Schwabe, J.-H.: Schwingungstechnische Auslegung von Betonrohrfertigern. Dissertation, Technische Universität Chemnitz, 2002
[4.5] Schwabe, J.-H.: Trends bei der Herstellung von Rohren und Schachtbauteilen. Betonwerk International Vol. 5/2003, 190-197
[4.6] Schwabe, J.-H.: Herstellung korrosionsbeständiger Beton- und Stahlbetonrohre. In: Tagungsband der 52. BetonTage Neu-Ulm (2008), 186-187
[4.7] Schwabe, J.-H.; Schulze, R.: Runde Schalungen mit Außenvibratoren – Analyse und Optimierung des Schwingungsverhaltens. In: Betonwerk+Fertigteil-Technik Vol. 11/2008, 18-25

[4.8] Zanker, G.: Fertigungsverfahren für Beton- und Stahlbetonrohre sowie Schacht-Bauteile. In: Betonwerk+Fertigteil-Technik Vol. 04/1989, 85-91

Chapter 5

[5.1] Autorenkollektiv: Neue Generation von Vibrationsformensystemen. Forschungsbericht, IFF Weimar e.V., 1999

[5.2] Kaysser, D.: Studie zur Analyse und Bewertung bekannter Fertigbausysteme hinsichtlich der Umweltrelevanz in der Phase der Fertigung; Forschungsbericht zum Projekt PRODOMO „Produktionsintegrierter Umweltschutz im Bereich des Hochbaus der Beton- und Fertigteilindustrie"; Forschungsbericht IFF Weimar e.V., 1997

[5.3] Autorenkollektiv: Handbuch Betonfertigteile, Betonwerkstein, Terrazzo. Verlag Bau+Technik GmbH, Düsseldorf 1999

[5.4] Kuch, H.; Palzer, U.; Schwabe, J.-H.: Formgebung und Verdichtung von Betonfertigteilen. Stand der Forschung und Entwicklung technischer Lösungen. BFT Betonwerk+ Fertigteil-Technik Vol. 11/2008

[5.5] Kuch, H.; Schwabe J.-H.: Verdichtungstechnologie für Betonfertigteile; Maschinendynamik und Messtechnik. BFT Betonwerk+ Fertigteiltechnik Vol. 08/1997

[5.6] Kuch, H.; Martin, J.; Schwabe, J.-H.; Beschleunigungsverteilungen an Vibrationsformen. BFT Betonwerk+ Fertigteil-Technik, Vol. 08/1999

[5.7] Kuch, H. et al.: Verdichtungskenngrößen bei Niederfrequenz-Einwirkung. Schriftenreihe der Forschungsvereinigung Bau- und Baustoffmaschinen, Vol. 24. December 2003

[5.8] Autorenkollektiv: Neue Generation von Vibrationsformensystemen. Institut für Fertigteiltechnik und Fertigbau Weimar e.V., 1999

[5.9] Karutz, H.: Maschinen und Anlagen für die Produktion von Spannbetonfertigdecken – BAUMA – Nachschau. BFT Betonwerk+Fertigteil-Technik Vol. 12/2004, 32-37

[5.10] Karutz, H.: Der X-Former – innovatives Konzept für Spannbetonfertigdecken. BFT Betonwerk+Fertigteil-Technik Vol. 04/2004, 34-41

[5.11] Schwarz, S.: Spannbetonhohlplatten- und doppelschalige Wandelementfertigung für internationalen Markt. BFT Betonwerk-Fertigteil-Technik Vol. 10/1998, 73-80

[5.12] Vollert/Weckenmann: Fragebogen für Anlagen zur Herstellung von Betonfertigteilen

The standard work on cement